SAFETY SYMBOLS

SAFETY SYMBOLS	HAZARD	EXAMPLES	PRECAUTION	REMEDY
DISPOSAL	Special disposal procedures need to be followed.	certain chemicals, living organisms	Do not dispose of these materials in the sink or trash can.	Dispose of wastes as directed by your teacher.
BIOLOGICAL	Organisms or other biological materials that might be harmful to humans	bacteria, fungi, blood, unpreserved tissues, plant materials	Avoid skin contact with these materials. Wear mask or gloves.	Notify your teacher if you suspect contact with material. Wash hands thoroughly.
EXTREME TEMPERATURE	Objects that can burn skin by being too cold or too hot	boiling liquids, hot plates, dry ice, liquid nitrogen	Use proper protection when handling.	Go to your teacher for first aid.
SHARP OBJECT	Use of tools or glassware that can easily puncture or slice skin	razor blades, pins, scalpels, pointed tools, dissecting probes, broken glass	Practice common-sense behavior and follow guidelines for use of the tool.	Go to your teacher for first aid.
FUME	Possible danger to respiratory tract from fumes	ammonia, acetone, nail polish remover, heated sulfur, moth balls	Make sure there is good ventilation. Never smell fumes directly. Wear a mask.	Leave foul area and notify your teacher immediately.
ELECTRICAL	Possible danger from electrical shock or burn	improper grounding, liquid spills, short circuits, exposed wires	Double-check setup with teacher. Check condition of wires and apparatus.	Do not attempt to fix electrical problems. Notify your teacher immediately.
IRRITANT	Substances that can irritate the skin or mucous membranes of the respiratory tract	pollen, moth balls, steel wool, fiberglass, potassium permanganate	Wear dust mask and gloves. Practice extra care when handling these materials.	Go to your teacher for first aid.
CHEMICAL	Chemicals that can react with and destroy tissue and other materials	bleaches such as hydrogen peroxide; acids such as sulfuric acid, hydrochloric acid; bases such as ammonia, sodium hydroxide	Wear goggles, gloves, and an apron.	Immediately flush the affected area with water and notify your teacher.
TOXIC	Substance may be poisonous if touched, inhaled, or swallowed	mercury, many metal compounds, iodine, poinsettia plant parts	Follow your teacher's instructions.	Always wash hands thoroughly after use. Go to your teacher for first aid.
OPEN FLAME	Open flame may ignite flammable chemicals, loose clothing, or hair	alcohol, kerosene, potassium permanganate, hair, clothing	Tie back hair. Avoid wearing loose clothing. Avoid open flames when using flammable chemicals. Be aware of locations of fire safety equipment.	Notify your teacher immediately. Use fire safety equipment if applicable.

 Eye Safety Proper eye protection should be worn at all times by anyone performing or observing science activities.

 Clothing Protection This symbol appears when substances could stain or burn clothing.

 Animal Safety This symbol appears when safety of animals and students must be ensured.

 Radioactivity This symbol appears when radioactive materials are used.

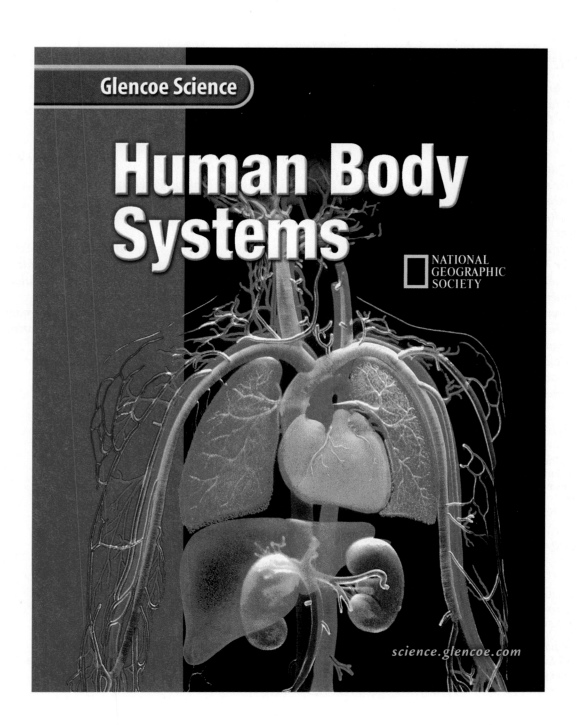

Glencoe Science

Human Body Systems

NATIONAL
GEOGRAPHIC
SOCIETY

science.glencoe.com

Glencoe
McGraw-Hill

New York, New York   Columbus, Ohio   Woodland Hills, California   Peoria, Illinois

Glencoe Science

Human Body Systems

Student Edition
Teacher Wraparound Edition
Interactive Teacher Edition CD-ROM
Interactive Lesson Planner CD-ROM
Lesson Plans
Content Outline for Teaching
Dinah Zike's Teaching Science with Foldables
Directed Reading for Content Mastery
Foldables: Reading and Study Skills
Assessment
   Chapter Review
   Chapter Tests
   ExamView Pro Test Bank Software
   Assessment Transparencies
   Performance Assessment in the Science Classroom
   The Princeton Review Standardized Test Practice Booklet
Directed Reading for Content Mastery in Spanish
Spanish Resources
English/Spanish Guided Reading Audio Program
Reinforcement

Enrichment
Activity Worksheets
Section Focus Transparencies
Teaching Transparencies
Laboratory Activities
Science Inquiry Labs
Critical Thinking/Problem Solving
Reading and Writing Skill Activities
Mathematics Skill Activities
Cultural Diversity
Laboratory Management and Safety in the Science Classroom
MindJogger Videoquizzes and Teacher Guide
Interactive CD-ROM with Presentation Builder
Vocabulary PuzzleMaker Software
Cooperative Learning in the Science Classroom
Environmental Issues in the Science Classroom
Home and Community Involvement
Using the Internet in the Science Classroom

"Study Tip," "Test-Taking Tip," and the "Test Practice" features in this book were written by The Princeton Review, the nation's leader in test preparation. Through its association with McGraw-Hill, The Princeton Review offers the best way to help students excel on standardized assessments.

The Princeton Review is not affiliated with Princeton University or Educational Testing Service.

Glencoe/McGraw-Hill

A Division of The McGraw-Hill Companies

Cover Images: Parts of several human body systems are shown here.

Send all inquiries to:
Glencoe/McGraw-Hill
8787 Orion Place
Columbus, OH 43240

ISBN 0-07-825574-0
Printed in the United States of America.
4 5 6 7 8 9 10 027/043 06 05 04 03 02

Authors

National Geographic Society
Education Division
Washington, D.C.

Edward Ortleb
Science Consultant
St. Louis Public Schools
St. Louis, Missouri

Dinah Zike
Educational Consultant
Dinah-Might Activities, Inc.
San Antonio, Texas

Content Consultants

Connie Rizzo, MD
Professor of Biology
Pace University
New York, New York

Series Safety Consultants

Malcolm Cheney, PhD
OSHA Chemical
Safety Officer
Hall High School
West Hartford, Connecticut

Aileen Duc, PhD
Science II Teacher
Hendrick Middle School
Plano, Texas

Sandra West, PhD
Associate Professor
of Biology
Southwest Texas State
University
San Marcos, Texas

Series Math Consultants

Michael Hopper, D. Eng
Manager of Aircraft Certification
Raytheon Company
Greenville, Texas

Teri Willard, EdD
Department of Mathematics
Montana State University
Belgrade, Montana

Reading Consultant

Elizabeth Babich
Special Education Teacher
Mashpee Public Schools
Mashpee, Massachusetts

Michelle Bailey
Northwood Middle School
Houston, Texas

Amy Morgan
Berry Middle School
Hoover, Alabama

Maureen Barrett
Thomas E. Harrington Middle School
Mt. Laurel, New Jersey

Billye Robbins
Jackson Intermediate
Pasadena, Texas

Janice Bowman
Coke R. Stevenson Middle School
San Antonio, Texas

Delores Stout
Bellefonte Middle School
Bellefonte, Pennsylvania

Cory Fish
Burkholder Middle School
Henderson, Nevada

Darcy Vetro-Ravndal
Middleton Middle School of Technology
Tampa, Florida

Tammy Ingraham
Westover Park Intermediate School
Canyon, Texas

Series Activity Testers

José Luis Alvarez, PhD
Math/Science Mentor Teacher
El Paso, Texas

Mary Helen Mariscal-Cholka
Science Teacher
William D. Slider Middle School
El Paso, Texas

Nerma Coats Henderson
Teacher
Pickerington Jr. High School
Pickerington, Ohio

José Alberto Marquez
TEKS for Leaders Trainer
El Paso, Texas

Science Kit and Boreal Laboratories
Tonawanda, New York

Nature of Science: Human Genome—2

CHAPTER 1

Structure and Movement—6

SECTION 1 **The Skeletal System** 8

SECTION 2 **The Muscular System** 14

NATIONAL GEOGRAPHIC Visualizing Human Body Levers. 16

SECTION 3 **The Skin** 20

Activity Measuring Skin Surface 25

Activity: Use the Internet
Similar Skeletons. 26

Oops Accidents in SCIENCE First Aid Dolls 28

CHAPTER 2

Nutrients and Digestion—34

SECTION 1 **Nutrition**. 36

NATIONAL GEOGRAPHIC Visualizing Vitamins 41

Activity Identifying Vitamin C Content 46

SECTION 2 **The Digestive System**. 47

Activity Particle Size and Absorption 54

TIME *Science and Society*
Eating Well 56

CHAPTER 3

Circulation—62

SECTION 1 **The Circulatory System** 64

NATIONAL GEOGRAPHIC Visualizing Atherosclerosis 70

Activity The Heart as a Pump 73

SECTION 2 **Blood** 74

SECTION 3 **The Lymphatic System** 80

Activity: Design Your Own Experiment
Blood Type Reactions. 82

TIME *Science and History*
Have a Heart 84

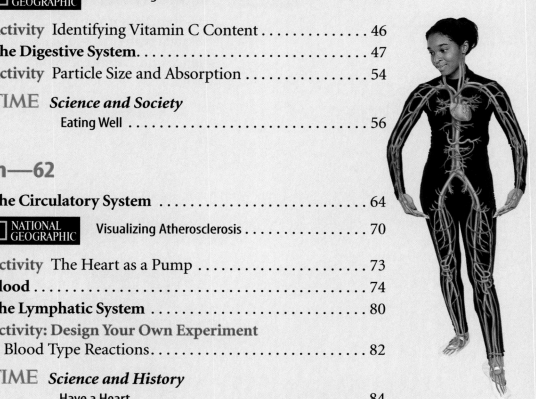

CHAPTER 4

Respiration and Excretion—90

SECTION 1 **The Respiratory System** 92

NATIONAL GEOGRAPHIC Visualizing Abdominal Thrusts 97

SECTION 2 **The Excretory System** 101

Activity Kidney Structure 107

Activity: Model and Invent
Simulating the Abdominal Thrust Maneuver 108

TIME *Science and History*
Overcoming the Odds 110

CHAPTER 5

Control and Coordination—116

SECTION 1 **The Nervous System** 118

NATIONAL GEOGRAPHIC Visualizing Nerve Impulse Pathways 120

Activity Improving Reaction Time 127

SECTION 2 **The Senses** 128

Activity: Design Your Own Experiment
Skin Sensitivity 136

Science and Language Arts
Sula 138

CHAPTER 6

Regulation and Reproduction—144

SECTION 1 **The Endocrine System** 146

NATIONAL GEOGRAPHIC Visualizing the Endocrine System 148

SECTION 2 **The Reproductive System** 151

Activity Interpreting Diagrams 156

SECTION 3 **Human Life Stages** 157

Activity Changing Body Proportions 166

Science Stats Facts About Infants 168

CHAPTER 7

Immunity and Disease—174

SECTION 1 **The Immune System** 176

SECTION 2 **Infectious Diseases** 181

NATIONAL GEOGRAPHIC Visualizing Koch's Rules 183

Activity Microorganisms and Disease 189

SECTION 3 **Noninfectious Diseases** 190

Activity: Design Your Own Experiment
Defensive Saliva 196

Science Stats Battling Bacteria 198

Field Guide

Emergencies Field Guide 208

Skill Handbooks—212

Reference Handbooks

A. Care and Use of a Microscope 237

B. Safety in the Science Classroom 238

C. SI/Metric to English, English to Metric Conversions . . 239

D. Diversity of Life 240

English Glossary—244

Spanish Glossary—250

Index—257

Interdisciplinary Connections/Activities

NATIONAL GEOGRAPHIC VISUALIZING

1 Human Body Levers 16
2 Vitamins 41
3 Atherosclerosis 70
4 Abdominal Thrusts 97
5 Nerve Impulse Pathways 120
6 The Endocrine System 148
7 Koch's Rules 183

TIME SCIENCE AND Society

2 **Science and Society:** Eating Well . . . 56

TIME SCIENCE AND HISTORY

3 **Science and History:** Have a Heart . . 84
4 **Science and History:** Overcoming
   the Odds 110

Accidents in SCIENCE

1 First Aid Dolls 28

Science Stats

6 Facts About Infants 168
7 Battling Bacteria 198

Science and Language Arts

5 Sula 138

Full Period Labs

1 Measuring Skin Surface 25
   Use the Internet: Similar Skeletons . . 26
2 Identifying Vitamin C Content 46
   Particle Size and Absorption 54
3 The Heart as a Pump 73
   Design Your Own Experiment:
      Blood Type Reactions 82
4 Kidney Structure 107
   Model and Invent: Simulating
      the Abdominal Thrust 108
5 Improving Reaction Time 127

Design Your Own Experiment:
   Skin Sensitivity 136
6 Interpreting Diagrams 156
   Changing Body Proportions 166
7 Microorganisms and Disease 189
Design Your Own Experiment:
   Defensive Saliva 196

Mini LAB

1 **Try at Home:** Comparing
      Muscle Activity 18
   Recognizing Why You Sweat 22
2 Comparing the Fat Content
      of Foods 39
   Try at Home: Modeling Absorption
      in the Small Intestine 52
3 **Try at Home:** Inferring How
      Hard the Heart Works 65
   Modeling Scab Formation 76
4 **Try at Home:** Comparing
      Surface Area 96
   Modeling Kidney Function 103
5 **Try at Home:** Observing Balance
      Control 132
   Comparing Sense of Smell 134
6 Graphing Hormone Levels 154
   Try at Home: Interpreting
      Fetal Development 160
7 **Try at Home:** Determining
      Reproduction Rates 179
   Observing Antiseptic Action 184

EXPLORE ACTIVITY

1 Observe muscle action 7
2 Model the digestive tract 35
3 Recognize transportation 63
4 Measure breathing rate 91
5 Observe a response 117
6 Model a chemical message 145
7 Model the spread of disease-
      causing organisms 175

Problem-Solving Activities

1 Is it unhealthful to snack
   between meals? 40
3 Will there be enough
   blood donors? 78
4 How does your body gain
   and lose water?. 104
7 Has the annual percentage of
   deaths from major diseases
   changed?..................... 185

Math Skills Activities

1 Estimating the Volume of Bones ... 11
5 Calculating Distance Using
   the Speed of Sound 133
6 Calculating Blood Sugar
   Percentage................... 147

Skill Builder Activities

Science
Communicating: 19, 53, 81, 100, 135
Concept Mapping: 19, 24, 72, 81, 106,
   126, 155
Interpreting Data: 45, 79
Making and Using Graphs: 188
Making and Using Tables: 13, 135, 195
Making Models: 165, 180
Predicting: 150
Recognizing Cause and Effect: 53, 188
Researching Information: 100, 150

Math
Solving One-Step Equations: 24, 106, 155
Using Percentages: 79

Technology
Using a Database: 72, 195
Using an Electronic Spreadsheet: 45, 165
Using a Word Processor: 126, 180
Using Graphics Software: 13

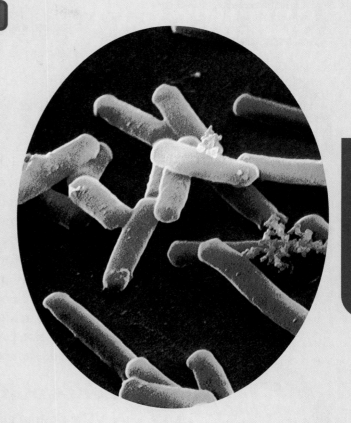

Science
INTEGRATION

Astronomy: 130
Chemistry: 21, 78, 122
Earth Science: 23, 43, 93, 105, 147, 182
Environmental Science: 53, 193
Physics: 69, 164

SCIENCE
Online

Research: 10, 15, 38, 50, 71, 75, 95, 98, 123, 125, 133, 153, 161, 178
Data Update: 81, 187

THE
PRINCETON
REVIEW

Test Practice: 33, 61, 89, 115, 143, 173, 203

Human Genome

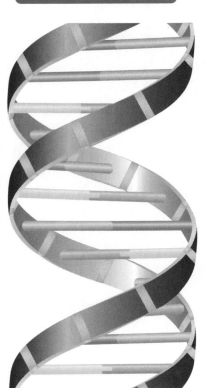

Figure 1
DNA contains two types of paired bases, adenine-thymine and cytosine-guanine.

By applying scientific methods and using technology, scientists completed the task of mapping the human genome. All of the DNA in an organism make up its genome. Although knowing the human genome allows for the possibilities of earlier diagnosis, better treatments, and even cures for many types of disorders, it also brings with it many questions about ethics and social values that cannot be answered by science. This feature presents information about the scientific achievements involved in sequencing the human genome. It also presents some questions raised by people in different fields that require careful consideration.

Genes and DNA

The human genome has approximately 30,000 genes. Genes are made of a complex chemical called DNA. In DNA there are four different substances called bases that only occur in two types of pairs. But, the number of paired bases and their order is unique for each species.

Before the 1950s, scientists could only look at a human cell's nucleus under a microscope and try to count the number of gene-containing chromosomes in it. But, in 2000, scientists finally determined the order of the 3 billion DNA bases on all the human chromosomes.

Figure 2
In 1983, Dr. Barbara McClintock received a Nobel Prize for discovering "jumping genes."

Unraveling the Code

How did scientists determine such a complex and lengthy sequence? Many discoveries about DNA were made before the order of the bases could be determined.

Dr. Barbara McClintock made one such discovery. She first recognized jumping DNA—stretches of DNA that can move around and between chromosomes—in Indian corn in the 1940s. In the 1980s, other scientists confirmed her findings. Scientists now hypothesize that this jumping DNA might be related to diseases such as hemophilia, leukemia, and breast cancer.

Figure 3
Sequencing small segments of DNA was one part of the Human Genome Project.

The Human Genome Project

An international effort to determine the human genome began in October of 1990. For many years, scientists did not have the technology to study chromosomes at the DNA level. In the 1970s and 1980s, computers were improved so that large amounts of data could be stored in small amounts of space. It takes three gigabytes of computer memory to store one human genome. This does not include additional information about the genome, only the order of bases.

DNA Sequencing

To determine the order of bases on a chromosome—a chromosome may be up to 250 million bases long—scientists sequence the DNA. First, chromosomes are broken into shorter pieces. Then, the fragments are analyzed to determine the bases. Finally, a computer is used to assemble the short sequences into long stretches, look for errors, and find other information.

What is science?

Science is concerned only with ideas or hypotheses that can be tested. Test results can be considered useful only if they are observable and repeatable. For scientists to learn whether an idea is correct or not, there must be observations or experiments that can show the idea to be true or false. For example, to make a working draft of the DNA sequence on the human genome, scientists identified 90 percent of the genes on each chromosome. Other scientists checked this information. The DNA sequence was not accepted until other scientists repeated it many times. Even before a working draft of the DNA sequence was completed, scientists checked their results multiple times.

Figure 4
This computer display shows some of the sequence results of The Human Genome Project.

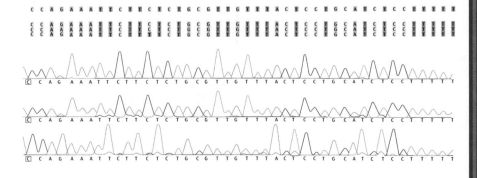

Better Science

Scientists are always striving for better experiments and more accurate observations that will increase their understanding. This means that scientific knowledge may change as scientists learn more. Recognition that the working draft of the human genome contained gaps and errors was an example of this. Scientific knowledge is the most reliable information people currently have. Although scientific knowledge is dependable, it is not certain or eternal.

Figure 5
Using up-to-date equipment allows scientists to obtain better results.

Limits of Science

Accepted theories change as scientists learn more about the world. This book describes the human body, but there is no information about how a person should behave or how they should think about their body. Science is not qualified to teach morality or spirituality.

Knowing the entire sequence of the human genome can help people in many ways. It can improve diagnoses of diseases and lead to the development of new medicines. Gene therapy—altering an organism's genes—may someday be used to treat or prevent disease. Science is the most reliable way of acquiring objective knowledge about the world, but it is not the only way. There are certain kinds of knowledge that science cannot uncover and some questions that are too complex for science alone to answer.

Technological Limits

Sometimes science cannot answer a question or solve a problem because scientists do not have the necessary tools or skills. Recall that scientists could not sequence the human genome until a complete understanding of DNA was learned and the computer technology to store the information was available. Many questions that scientists cannot answer today may be answered in the future as new technology is developed.

Figure 6
These are some of the computers used to sequence the human genome.

What Science Can't Answer

Even if there were no technical limits, science still could answer only certain kinds of questions. The questions that science can answer are those about facts—about the way things are in the world. But science cannot answer questions about values—how things "should be". Scientific discoveries can raise questions about values. Human genome research, for example, has raised many ethical, legal, and social issues. Some questions raised by this research are:

- Who will be able to find out about a person's genome and how will the information be used?
- Should genetic testing for a disease be performed when no treatment is available?
- How will knowledge that someone may develop a genetic disease affect that person? How will society regard such an individual?
- Do genes make people behave in certain ways? Can they always control their behavior?

Science may provide information that can help people understand issues better, but people have to make their own decisions based on their own values and beliefs.

Figure 7
Many decisions must be made about the application of information gained from the Human Genome Project.

In this book, you will learn about human body systems. Some of this information has been known for centuries. Other information is from recent discoveries, such as the understanding of genetic links to certain disorders. Gene therapy is a way to treat, cure, or prevent genetically linked disease by altering a person's genes. Today, research in gene therapy is just beginning. But someday it may be available to help persons with genetic diseases. Research this topic and debate it with your classmates. Consider such questions as, "What is normal and what is a disorder? Who decides? Are disabilities diseases? Do they need to be cured or prevented?" Early attempts at gene therapy will be very expensive. "Who will pay for the therapies? Who will get these therapies?"

Structure and Movement

A calm, misty morning's silence is broken when a huge, snarling *Tyrannosaurus rex* rushes to pounce on its prey. Without an internal skeleton, the *T. rex* would be a large, formless mass of flesh. Movement would be impossible. Its internal organs would not be protected. In this chapter, learn about the internal structures that support you.

What do you think?

Science Journal Look at the picture below with a classmate. Discuss what you think this might be or what is happening. Here's a hint: *They help you avoid dangerous situations.* Write your answer or best guess in your Science Journal.

The expression "Many hands make light work" is also true when it comes to muscles in your body. In fact, hundreds of muscles and bones work together to bring about smooth, easy movement. Muscle interactions enable you to pick up a penny or lift a 10-kg weight. Try the following activity to see and feel how your muscles work in pairs.

Observe muscle action

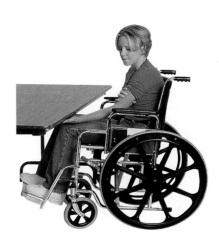

1. Sit on a chair at an empty table and place the palm of one hand under the edge of the table.
2. Push your hand up against the table. Do not push too hard.
3. Use your other hand to feel the muscles located on both sides of your upper arm, as shown in the photo.
4. Next, place your palm on the top of the table and push down. Again, feel the muscles in your upper arm.

Observe

Which muscles felt harder when pushing up on the table? When pushing down on the table? Describe in your Science Journal how the different muscles in your upper arm were working during each movement.

Before You Read

Making a Main Ideas Study Fold Make the following Foldable to help you identify the main topics about structure and movement.

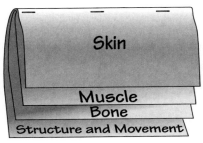

1. Stack two sheets of paper in front of you so the short side of both sheets is at the top.
2. Slide the top sheet up so that about 4 cm of the bottom sheet show.
3. Fold both sheets top to bottom to form four tabs and staple along the top fold as shown.
4. Label the flaps *Skin, Muscle, Bone,* and *Structure and Movement* as shown. Before you read the chapter, list the things you know about each on the back of each flap.
5. As you read the chapter, write about the changes that occur in the skin, muscle, and bone on their flap.

The Skeletal System

What **You'll Learn**

- **Identify** five functions of the skeletal system.
- **Compare and contrast** movable and immovable joints.

Vocabulary

skeletal system
periosteum
cartilage
joint
ligament

Why **It's Important**

You'll begin to understand how each of your body parts moves and what happens that allows you to move them.

Living Bones

Often in a horror movie, a mad scientist works frantically in his lab with a complete human skeleton hanging silently in the corner. When looking at a skeleton, you might think that bones are dead structures made of rocklike material. Although these bones are no longer living, the bones in your body are very much alive. Each is a living organ made of several different tissues. Like all the other living tissues in your body, bone tissue is made of cells that take in nutrients and use energy. Bone cells have the same needs as other body cells.

Functions of Your Skeletal System All the bones in your body make up your **skeletal system.** Like the framework of the building in **Figure 1,** the skeletal system is the framework of your body and has five major functions.

1. The skeleton gives shape and support to your body.

2. Bones protect your internal organs. For example, ribs surround the heart and lungs, and the skull encloses the brain.

3. Major muscles are attached to bone and help them move.

4. Blood cells are formed in the center of many bones in soft tissue called red marrow.

5. Major quantities of calcium and phosphorous compounds are stored in the skeleton for later use. Calcium and phosphorus make bones hard.

Figure 1
Nonliving structures—like cars, houses, and this building—require some type of framework for support, as does your body.

Bone Structure

Several characteristics of bones are noticeable. The most obvious, as seen in **Figure 2,** are the differences in their sizes and shapes. The shapes of bones are inherited. However, a bone's shape can change when the attached muscles are used.

Looking at bone through a magnifying glass will show you that it isn't smooth. Bones have bumps, edges, round ends, rough spots, and many pits and holes. Muscles and ligaments attach to some of the bumps and pits. In your body blood vessels and nerves enter and leave through the holes. Internal characteristics, how a bone looks from the inside, and external characteristics, how the same bone looks from the outside, are shown in **Figure 3.**

A living bone's surface is covered with a tough, tight-fitting membrane called the **periosteum** (pur ee AHS tee um). Small blood vessels in the periosteum carry nutrients into the bone. Cells involved in the growth and repair of bone also are found in the periosteum. Under the periosteum are two different types of bone tissue—compact bone and spongy bone.

Compact Bone Directly under the periosteum is a hard, strong layer called compact bone. Compact bone gives bones strength. It has a framework containing deposits of calcium phosphate. These deposits make the bone hard. Bone cells and blood vessels also are found in this layer. This framework is living tissue and even though it's hard, it keeps bone from being too rigid, brittle, or easily broken.

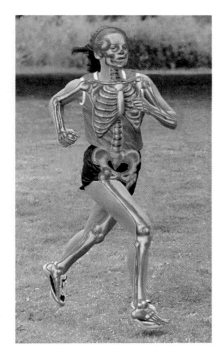

Figure 2
The 206 bones of the body are connected, forming a framework called the skeleton.

Figure 3
Bone is made of layers of living tissue. Compact bone is arranged in circular structures called Haversian systems—tiny, connected channels through which blood vessels and nerve fibers pass.

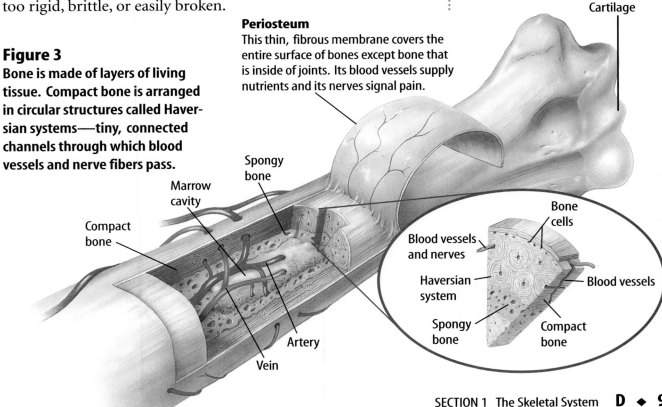

Periosteum
This thin, fibrous membrane covers the entire surface of bones except bone that is inside of joints. Its blood vessels supply nutrients and its nerves signal pain.

Cartilage

Spongy bone

Marrow cavity

Compact bone

Artery

Vein

Bone cells

Blood vessels and nerves

Haversian system

Spongy bone

Blood vessels

Compact bone

Research Visit the Glencoe Science Web site at **science.glencoe.com** for information on new techniques for treating bone fractures. Using captions, illustrate one technique in your Science Journal.

Spongy Bone Spongy bone is located toward the ends of long bones such as those in your thigh and upper arm. Spongy bone has many small, open spaces that make bones lightweight. If all your bones were completely solid, you'd have greater mass. In the centers of long bones are large openings called cavities. These cavities and the spaces in spongy bone are filled with a substance called marrow. Some marrow is yellow and is composed of fat cells. Red marrow produces red blood cells at an incredible rate of 2 million to 3 million cells per second.

Cartilage The ends of bones are covered with a smooth, slippery, thick layer of tissue called **cartilage.** Cartilage does not contain blood vessels or minerals. It is flexible and important in joints because it acts as a shock absorber. It also makes movement easier by reducing friction that would be caused by bones rubbing together. Cartilage can be damaged because of disease, injury, or years of use. People with damaged cartilage experience pain when they move.

✔ **Reading Check** *What is cartilage?*

Bone Formation

Although your bones have some hard features, they have not always been this way. Months before your birth, your skeleton was made of cartilage. Gradually the cartilage broke down and was replaced by bone, as illustrated in **Figure 4.** Bone-forming cells called osteoblasts (AHS tee oh blasts) deposit the minerals calcium and phosphorus in bones, making the bone tissue hard. At birth, your skeleton was made up of more than 300 bones. As you developed, some bones fused, or grew together, so that now you have only 206 bones.

Healthy bone tissue is always being formed and re-formed. Osteoblasts build up bone. Another type of bone cell, called an osteoclast, breaks down bone tissue in other areas of the bone. This is a normal process in a healthy person. When osteoclasts break bone down, they release calcium and phosphorus into the bloodstream. This process maintains the elements calcium and phosphorus in your blood at about the levels they need to be. These elements are necessary for the working of your body, including the movement of your muscles.

Figure 4
Cartilage is replaced slowly by bone as solid tissue grows outward. Over time, the bone reshapes to include blood vessels, nerves, and marrow. *What type of bone cell builds up bone?*

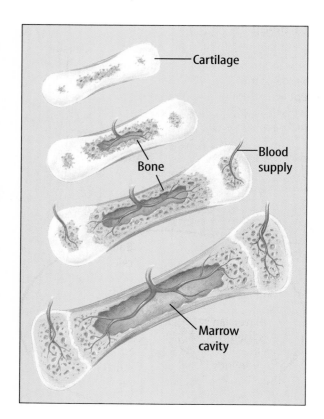

Cartilage

Bone

Blood supply

Marrow cavity

Joints

What will you do during your lunch break today? You may sit at a table, pick up a sandwich, bite off a piece of a carrot and chew it, or walk to class. All of these motions are possible because your skeleton has joints.

Any place where two or more bones come together is a **joint.** The bones in healthy joints are kept far enough apart by a thin layer of cartilage so that they do not rub against each other as they move. The bones are held in place at these joints by a tough band of tissue called a **ligament.** Many joints, such as your knee, are held together by more than one ligament. Muscles move bones by moving joints.

What would you do if you sprained your ankle? To find out more about what to do in emergency situations, see the **Emergencies Field Guide** at the back of the book.

Math Skills Activity

Estimating the Volume of Bones

Example Problem

The Haversian systems found in the cross section of your bones are arranged in long cylinders. This cylindrical shape allows your bones to withstand great pressure. Although bones are not perfectly shaped, many of them are cylinder shaped. Estimate the volume of a bone that is 36 cm long and is 7 cm in diameter.

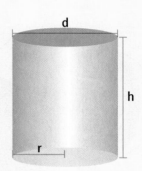

Solution

1. *This is what you know:* The bone has a shape of a cylinder whose height, *h*, measures 36 cm and whose diameter is 7.0 cm.

2. *This is what you want to find:* volume of the cylinder

3. *This is the equation you need to use:* Volume = π × (radius)^2 × height, or $V = \pi \times r^2 \times h$

   A radius is one-half the diameter ($\frac{1}{2}$ × 7), so r = 3.5, h = 36, and π = 3.14.

4. *Substitute the values for π, radius, and height into the equation and solve for V:* $V = 3.14 \times (3.5)^2 \times 36$
   $V = 1{,}384.74$
   The volume of the bone is 1,384.74 cm^3.

   To check your answer, first divide it by 3.14 and then divide that number by (3.5)^2. This number should be the height of the bone.

> **Practice Problem**
>
> Find the volume of a bone that has a height of 12 cm and a diameter of 2.4 cm.

Immovable Joints Refer to **Figure 5** as you learn about different types of joints. Joints are broadly classified as immovable or movable. An immovable joint allows little or no movement. The joints of the bones in your skull and pelvis are classified as immovable joints.

Movable Joints All movements, including somersaulting and working the controls of a video game, require movable joints. A movable joint allows the body to make a wide range of motions. There are several types of movable joints—pivot, ball and socket, hinge, and gliding. In a pivot joint, one bone rotates in a ring of another bone that does not move. Turning your head is an example of a pivot movement.

A ball-and-socket joint consists of a bone with a rounded end that fits into a cuplike cavity on another bone. A ball-and-socket joint provides a wider range of motion than a pivot joint does. That's why your legs and arms can swing in almost any direction.

A third type of joint is a hinge joint. This joint has a back-and-forth movement like hinges on a door. Elbows, knees, and fingers have hinge joints. Hinge joints have a smaller range of motion than the ball-and-socket joint. They are not dislocated as easily, or pulled apart, as a ball-and-socket joint can be.

A fourth type of joint is a gliding joint in which one part of a bone slides over another bone. Gliding joints also move in a back-and-forth motion and are found in your wrists and ankles and between vertebrae. Gliding joints are used the most in your body. You can't write a word, use a joy stick, or take a step without using a gliding joint.

Figure 5
When a basketball player shoots a ball, several types of joints are in action. *What other activities use several types of joints?*

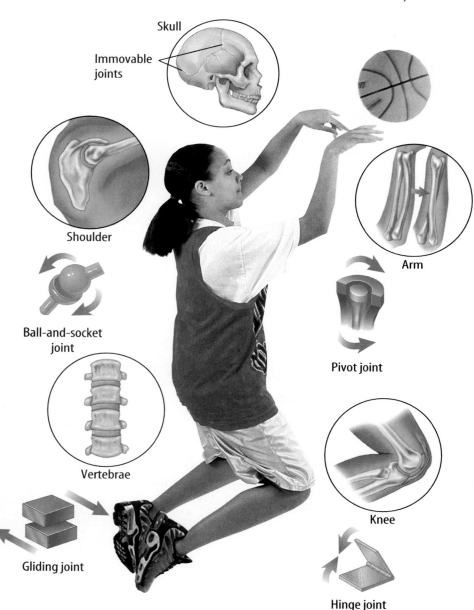

Skull

Immovable joints

Shoulder

Ball-and-socket joint

Vertebrae

Gliding joint

Arm

Pivot joint

Knee

Hinge joint

Moving Smoothly When you rub two pieces of chalk together, their surfaces begin to wear away, and they get reshaped. Without the protection of the cartilage at the end of your bones, they also would wear away at the joints. Cartilage helps make joint movement easier. It reduces friction and allows bones to slide more easily over each other. Shown in **Figure 6,** pads of cartilage, called disks, are located between the vertebrae in your back. They act as a cushion and prevent injury to your spinal cord. A fluid that comes from nearby blood vessels also lubricates the joint.

Your skeleton is a living framework of bones. Bones support the body and supply it with minerals and blood cells. Joints make the framework flexible.

Common Joint Problems Arthritis is the most common joint problem. The term *arthritis* describes more than 100 different diseases that can damage the joints. About one out of every seven people in the United States suffers from arthritis. All forms of arthritis begin with the same symptoms: pain, stiffness, and swelling of the joints.

Two common types of arthritis include osteoarthritis and rheumatoid arthritis. Older people often suffer from osteoarthritis, in which cartilage breaks down because of years of use. Rheumatoid arthritis occurs in both young and older adults. This disorder is an ongoing condition in which the body's immune system tries to destroy its own tissues.

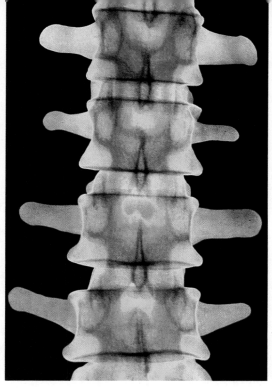

Figure 6
A colored X ray of the human backbone shows disks of cartilage between the vertebrae.

Section 1 Assessment

1. What are the five major functions of the skeletal system?
2. Name and give an example of both a movable joint and an immovable joint.
3. What are the functions of cartilage in your skeletal system?
4. What are ligaments?
5. **Think Critically** A thick band of bone forms around a broken bone as it heals. In time, the thickened band disappears. Explain how this extra bone can disappear over time.

Skill Builder Activities

6. **Making and Using Tables** Use a table to classify the bones of the human body as follows: *long, short, flat,* and *irregular.* **For more help, refer to the** Science Skill Handbook.

7. **Using Graphics Software** Using graphics software, make a circle graph that shows how an adult's bones are distributed: *29 skull bones, 26 vertebrae, 25 ribs, four shoulder bones, 60 arm and hand bones, two hip bones,* and *60 leg and feet bones.* **For more help, refer to the** Technology Skill Handbook.

The Muscular System

As You Read

What You'll Learn

- **Identify** the major function of the muscular system.
- **Compare and contrast** the three types of muscles.
- **Explain** how muscle action results in the movement of body parts.

Vocabulary

muscle
voluntary muscle
involuntary muscle
skeletal muscle
tendon
cardiac muscle
smooth muscle

Why It's Important

The muscular system is responsible for how you move and the production of heat in your body. Muscles also give your body its shape.

Movement of the Human Body

The golfer looks down the fairway and then at the golf ball. With intense concentration and muscle coordination, the golfer swings the club along a graceful arc and connects with the ball. The ball sails through the air, landing inches away from the flag. The crowd applauds. A few minutes later, the final putt is made and the tournament is won. The champion has learned how to use controlled muscle movement to bring success.

Muscles help make all of your daily movements possible. **Figure 7** shows which muscles connect some of the bones in your body. A **muscle** is an organ that can relax, contract, and provide the force to move your body parts. In the process, energy is used and work is done. Imagine how much energy the more than 600 muscles in your body use each day. No matter how still you might try to be, some muscles in your body are always moving. You're breathing, your heart is beating, and your digestive system is working.

Figure 7
Your muscles come in many shapes and sizes. Even simple movements require the coordinated use of several muscles. The muscles shown here are only those located directly under the skin. Beneath these muscles are middle and deep layers of muscles.

A

B

Muscle Control Your hand, arm, and leg muscles are voluntary. So are the muscles of your face, shown in **Figure 8.** You can choose to move them or not to move them. Muscles that you are able to control are called **voluntary muscles.** In contrast, **involuntary muscles** are muscles you can't control consciously. They go on working all day long, all your life. Blood gets pumped through blood vessels, and food is moved through your digestive system by the action of involuntary muscles.

 **Reading Check** *What is another body activity that is controlled by involuntary muscles?*

Your Body's Simple Machines—Levers

Physics INTEGRATION Your skeletal system and muscular system work together when you move, in the same way that the parts of a bicycle work together when it moves. A machine, such as a bicycle, is any device that makes work easier. A simple machine does work with only one movement, like a hammer. The hammer is a type of simple machine called a lever, which is a rod or plank that pivots or turns about a point. This point is called a fulcrum. The action of muscles, bones, and joints working together is like a lever. In your body, bones are rods, joints are fulcrums, and contraction and relaxation of muscles provide the force to move body parts. Levers are classified into three types—first class, second class, and third class. Examples of the three types of levers that are found in the human body are shown in **Figure 9.**

Figure 8
Facial expressions generally are controlled by voluntary muscles. It takes only 13 muscles to smile, but 43 muscles to frown.

Research Visit the Glencoe Science Web site at **science.glencoe.com** for recent news or magazine articles about replacing diseased joints. Communicate to your class what you learn.

Figure 9

All three types of levers—first-class, second-class, and third-class—are found in the human body. In the photo below, a tennis player prepares to serve a ball. As shown in the accompanying diagrams, the tennis player's stance demonstrates the operation of all three classes of levers in the human body.

▲ Fulcrum
▼ Effort force
■ Load

FIRST-CLASS LEVER
The fulcrum lies between the effort force and the load. This happens when the tennis player uses his neck muscles to tilt his head back.

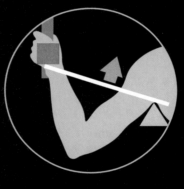

THIRD-CLASS LEVER
The effort force is between the fulcrum and the load. This happens when the tennis player flexes the muscles in his arm and shoulder.

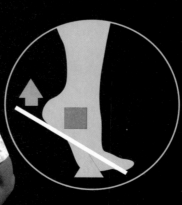

SECOND-CLASS LEVER
The load lies between the fulcrum and the effort force. This happens when the tennis player's calf muscles lift the weight of his body up on his toes.

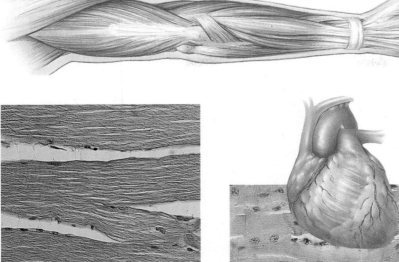

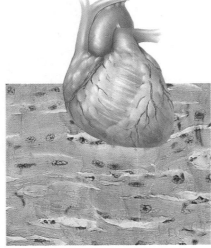

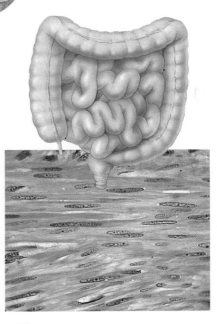

A Skeletal muscles move bones. The muscle tissue appears striped, or striated, and is attached to bone.

B Cardiac muscle is found only in the heart. The muscle tissue has striations.

C Smooth muscle is found in many of your internal organs, such as the digestive tract. This muscle tissue is nonstriated.

Classification of Muscle Tissue

All the muscle tissue in your body is not the same. The three types of muscles are skeletal, smooth, and cardiac. **Skeletal muscles** are the muscles that move bones. They are more common than other muscle types and are attached to bones by thick bands of tissue called **tendons.** When viewed under a microscope, skeletal muscle cells look striped, or striated (STRI ayt ud). You can see the striations in **Figure 10A.** Skeletal muscles are voluntary muscles. You choose when to walk or when not to walk. Skeletal muscles tend to contract quickly and tire more easily than involuntary muscles do.

The remaining two types of muscles are shown in **Figures 10B** and **10C. Cardiac muscle** is found only in the heart. Like skeletal muscle, cardiac muscle is striated. This type of muscle contracts about 70 times per minute every day of your life. **Smooth muscles** are found in your intestines, bladder, blood vessels, and other internal organs. They are nonstriated, involuntary muscles that slowly contract and relax. Internal organs are made of one or more layers of smooth muscles.

Figure 10
The three types of muscle tissue are skeletal muscle, cardiac muscle, and smooth muscle.

Figure 11

A When the flexor (hamstring) muscles of your thigh contract, the lower leg is brought toward the thigh. **B** When the extensor (quadriceps) muscles contract, the lower leg is straightened.

What class of lever is shown here?

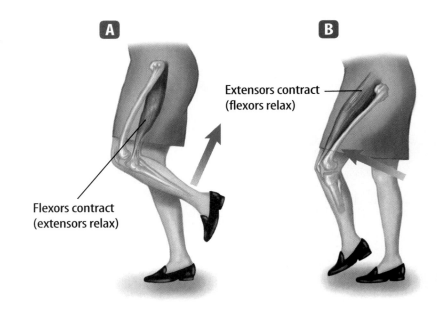

Extensors contract (flexors relax)

Flexors contract (extensors relax)

Comparing Muscle Activity

Procedure

1. Hold a light **book** in your outstretched hand over a dining or kitchen **table.**
2. Lift the book from this position to a height of 30 cm from the table 20 times.

Analysis

1. Compare your arm muscle activity to the continuous muscle activity of the heart.
2. Infer whether heart muscles become tired.

Working Muscles

How do muscles allow you to move your body? You move because pairs of skeletal muscles work together. When one muscle of a pair contracts, the other muscle relaxes, or returns to its original length, as shown in **Figure 11.** Muscles always pull. They never push. When the muscles on the back of your upper leg contract, they shorten and pull your lower leg back and up. When you straighten your leg, the back muscles lengthen and relax, and the muscles on the front of your upper leg contract. Compare how the muscles of your legs work with how the muscles of your arms work.

Changes in Muscles Over a period of time, muscles can become larger or smaller, depending on whether or not they are used. Skeletal muscles that do a lot of work, such as those in your writing hand, become strong and large. For example, many soccer and basketball players have noticeably larger, defined leg muscles. Muscles that are given regular exercise respond quickly to stimuli. Some of this change in muscle size is because of an increase in the number of muscle cells. However, most of this change in muscle size is because individual muscle cells become larger.

In contrast, if you participate only in nonactive pastimes such as watching television or playing computer games, your muscles will become soft and flabby and will lack strength. Muscles that aren't exercised become smaller in size. When someone is paralyzed, his or her muscles become smaller due to lack of exercise.

How Muscles Move Your muscles need energy to contract and relax. Your blood carries energy-rich molecules to your muscle cells where the chemical energy stored in these molecules is released. As the muscle contracts, this released energy changes to mechanical energy (movement) and thermal energy (heat), as shown in **Figure 12.** When the supply of energy-rich molecules in a muscle is used up, the muscle becomes tired and needs to rest. During this resting period, your blood supplies more energy-rich molecules to your muscle cells. The heat produced by muscle contractions helps keep your body temperature constant.

Figure 12
Chemical energy is needed for muscle activity. During activity, chemical energy supplied by food is changed into mechanical energy (movement) and thermal energy (heat).

Section 2 Assessment

1. What is the function of muscles?
2. Compare and contrast the three types of muscle tissue.
3. What type of muscle tissue is found in your heart?
4. Describe how a muscle attaches to a bone.
5. **Think Critically** What happens to your upper arm muscles when you bend your arm at the elbow?

Skill Builder Activities

6. **Concept Mapping** Using a concept map, sequence the activities that take place when you bend your leg at the knee. **For more help, refer to the** Science Skill Handbook.
7. **Communicating** Write a paragraph in your Science Journal about the three forms of energy involved in a muscle contraction. **For more help, refer to the** Science Skill Handbook.

The Skin

Your Largest Organ

What is the largest organ in your body? When you think of an organ, you might imagine your heart, stomach, lungs, or brain. However, your skin is the largest organ of your body. Much of the information you receive about your environment comes through your skin. You can think of your skin as your largest sense organ.

Skin Structures

Skin is made up of three layers of tissue—the epidermis, the dermis, and a fatty layer—as shown in **Figure 13.** Each layer of skin is made of different cell types. The **epidermis** is the outer, thinnest layer of your skin. The epidermis's outermost cells are dead and water repellent. Thousands of epidermal cells rub off every time you take a shower, shake hands, blow your nose, or scratch your elbow. New cells are produced constantly at the base of the epidermis. These new cells move up and eventually replace those that are rubbed off.

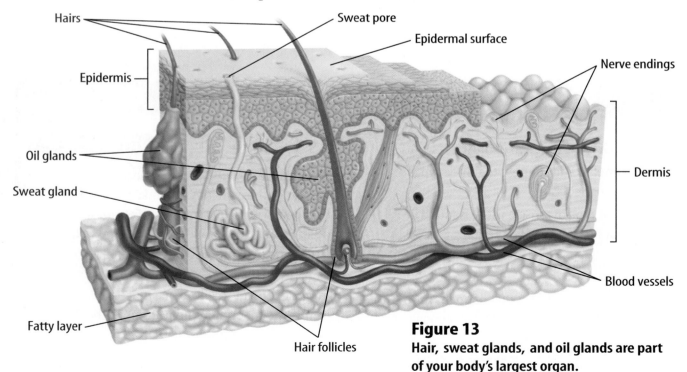

Figure 13
Hair, sweat glands, and oil glands are part of your body's largest organ.

Melanin Cells in the epidermis produce the chemical melanin (MEL uh nun). **Melanin** is a pigment that protects your skin and gives it color. The different amounts of melanin produced by cells result in differences in skin color, as shown in **Figure 14.** When your skin is exposed to ultraviolet rays, melanin production increases and your skin becomes darker. Lighter skin tones have less protection from the Sun. Such skin burns more easily and may be more susceptible to skin cancer.

Other Skin Layers The **dermis** is the layer of cells directly below the epidermis. This layer is thicker than the epidermis and contains many blood vessels, nerves, muscles, oil and sweat glands, and other structures. Below the dermis is a fatty region that insulates the body. This is where much of the fat is deposited when a person gains weight.

Skin Functions

Your skin is not only the largest organ of your body, it also carries out several major functions, including protection, sensory response, formation of vitamin D, regulation of body temperature, and ridding the body of wastes. The most important function of the skin is protection. The skin forms a protective covering over the body that prevents physical and chemical injury. Some bacteria and other disease-causing organisms cannot pass through the skin as long as it is unbroken. Glands in the skin secrete fluids that can damage or destroy some bacteria. The skin also slows down water loss from body tissues.

Specialized nerve cells in the skin detect and relay information to the brain, making the skin a sensory organ, too. Because of these cells, you are able to sense the softness of a cat, the sharpness of a pin, or the heat of a frying pan.

Earth Science
INTEGRATION

Research the effects of ultraviolet radiation on skin. Mountain climbers risk becoming severely sunburned even in freezing temperatures due to increased ultraviolet (UV) radiation. Why is UV radiation stronger on top of mountains? Record your answers in your Science Journal.

Figure 14
Melanin gives skin and eyes their color. The more melanin that is present, the darker the color is. This pigment provides protection from damage caused by harmful light energy.

Figure 15
Normal human body temperature is about 37°C. Temperature varies throughout the day. The highest body temperature is reached at about 11 A.M. and the lowest at around 4 A.M. At 43°C (109.5°F) fatal bleeding results, causing death.

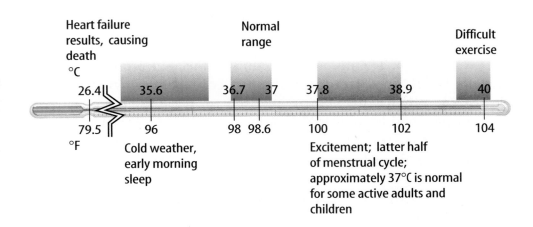

Heart failure results, causing death

Normal range

Difficult exercise

°C
26.4 35.6 36.7 37 37.8 38.9 40
79.5 96 98 98.6 100 102 104
°F

Cold weather, early morning sleep

Excitement; latter half of menstrual cycle; approximately 37°C is normal for some active adults and children

Vitamin D Formation Another important function of skin is the formation of vitamin D. Small amounts of this vitamin are produced in the presence of ultraviolet light from a fatlike molecule in your epidermis. Vitamin D is essential for good health because it helps your body absorb calcium into your blood from food in your digestive tract.

Heat and Waste Exchange Humans can withstand a limited range of body temperatures, as shown in **Figure 15.** Your skin plays an important role in regulating your body temperature. Blood vessels in the skin can help release or hold heat. If the blood vessels expand, or dilate, blood flow increases and heat is released. In contrast, less heat is released when the blood vessels constrict. Think of yourself after running—are you flushed red or pale and shivering?

The adult human dermis has about 3 million sweat glands. These glands help regulate the body's temperature and excrete wastes. When the blood vessels dilate, pores open in the skin that lead to the sweat glands. Perspiration, or sweat, moves out onto the skin. Heat transfers from the body to the sweat on the skin. Eventually, this sweat evaporates, removing the heat and cooling the skin. This system eliminates excess heat produced by muscle contractions.

✓ **Reading Check** *What are two functions of sweat glands?*

As your cells use nutrients for energy, they produce wastes. Such wastes, if not removed from your body, can act as poisons. In addition to helping regulate your body's temperature, sweat glands release water, salt, and other waste products. If too much water and salt are released by sweating during periods of extreme heat or physical exertion, you might feel light-headed or even faint.

Skin Injuries and Repair

Your skin often is bruised, scratched, burned, ripped, and exposed to harsh conditions like cold and dry air. In response, the skin produces new cells in its epidermis and repairs tears in the dermis. When the skin is injured, disease-causing organisms can enter the body rapidly. An infection often results.

Bruises Bruises are common, everyday events. Playing sports or working around your house often results in small injuries. What is a bruise and how does your body repair it?

When you have a bruise, your skin is not broken but the tiny blood vessels underneath the skin have burst. Red blood cells from these broken blood vessels leak into the surrounding tissue. These blood cells then break down, releasing a chemical called hemoglobin. The hemoglobin gradually breaks down into its components, called pigments. The color of these pigments causes the bruised area to turn shades of blue, red, and purple, as shown in **Figure 16.** Swelling also may occur. As the injury heals, the bruise eventually turns yellow as the pigment in the red blood cells is broken down even more and reenters the bloodstream. After all of the pigment is absorbed into the bloodstream, the bruise disappears and the skin looks normal again.

✔ Reading Check *What is the source of the yellow color of a bruise that is healing?*

Cuts Any tear in the skin is called a cut. Blood flows out of the cut until a clot forms over it. A scab then forms, preventing bacteria from entering the body. Cells in the surrounding blood vessels fight infection while the skin cells beneath the scab grow to fill the gap in the skin. In time, the scab falls off, leaving the new skin behind. If the cut is large enough, a scar may develop because of the large amounts of thick tissue fibers that form.

The body generally can repair bruises and small cuts. What happens when severe burns, some diseases, and surgeries result in injury to large areas of skin? Sometimes, not enough skin cells are left that can divide to replace this lost layer. If not treated, this can lead to rapid water loss from skin and muscle tissues, leading to infection and possible death. Skin grafts can prevent such problems. What are skin grafts?

Chemistry
INTEGRATION

Oil and sweat glands in your skin cause the skin to be acidic. With a pH between 3 and 5, the growth of potential disease-causing micro-organisms on your skin is reduced. What does pH mean? What common substances around your home have a pH value similar to that of your skin? Research to find these answers and then record them in your Science Journal.

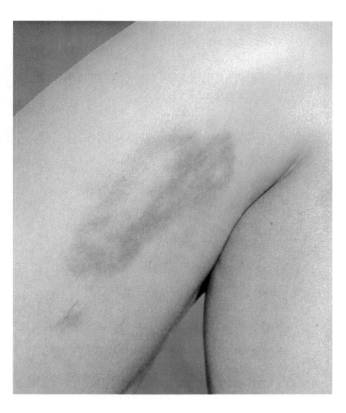

Figure 16
Bruising occurs when tiny blood vessels beneath the skin burst.

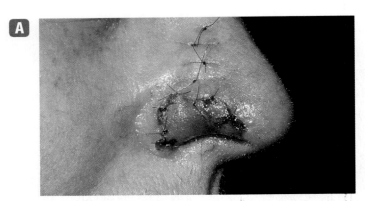

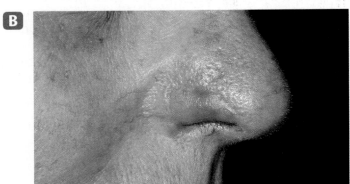

Skin Grafts Pieces of skin that are cut from one part of a person's body and then moved to the injured or burned area where there is no skin are called skin grafts. This new graft of skin is kept alive by nearby blood vessels and soon becomes part of the surrounding skin. Successful skin grafts, shown in **Figure 17,** must be taken from the victim's own body or possibly an identical twin. Skin transplants from other sources are rejected in about three weeks.

What can be done for severe burn victims who have little healthy skin left? Since the 1880s, doctors have used the skin of dead individuals, called cadavers, to treat such burns temporarily. However, the body usually rejects this skin, so the skin must be replaced continually until the burn heals.

A recent advancement in skin repair uses temporary grafts from cadavers to prevent immediate infections, while scientists grow large sheets of epidermis from small pieces of the burn victim's healthy skin. After 19 to 21 days, the cadaver skin patch is removed and the new epidermis is applied. With new technologies, severe cases of skin loss or damage that cannot be repaired may no longer be fatal.

Figure 17
A cancerous growth was removed from the nose of a 69-year-old woman. **A** A piece of skin removed from her scalp was grafted onto her nose to replace the lost skin. **B** The skin graft is healing after only one month.

Section Assessment

1. Compare and contrast the epidermis and dermis.
2. List the major functions of the body's largest organ, skin.
3. How does skin help prevent disease in the body?
4. Describe one way doctors are able to repair severe skin damage from burns, injuries, or surgeries.
5. **Think Critically** Why is a person who has been severely burned in danger of dying from loss of water?

Skill Builder Activities

6. **Concept Mapping** Make an events chain concept map to show how skin helps keep body temperature constant. **For more help, refer to the** Science Skill Handbook.
7. **Solving One-Step Equations** Your skin varies in thickness. The thin, almost transparent eyelids are 0.5 mm thick. On the palms of your hand and soles of your feet, skin is up to 0.4 cm thick. How many times thicker is the skin on the soles of your feet compared to your eyelids? **For more help, refer to the** Math Skill Handbook.

Activity

Measuring Skin Surface

Skin covers the entire surface of your body and is your body's largest organ. Skin cells make up a layer of skin about 2 mm thick. These cells are continually lost and re-formed. Skin cells are shed daily at a rate of an average of 50,000 cells per minute. In one year, humans lose about 2 kg of skin and hair. How big is this organ? Find the surface area of human skin.

What You'll Investigate
How much skin covers your body?

Goals
- **Estimate** the surface area of skin that covers the body of a middle-school student.

Materials
10 large sheets of newspaper
scissors
tape
meterstick or ruler

Safety Precautions 🥽 🖐

Procedure

1. Form groups of three or four, either all female or all male. Select one person from your group to measure the surface area of his or her skin.

2. **Estimate** how much skin covers the average student in your classroom. In your Science Journal, record your estimation.

3. Wrap newspaper snugly around each part of your classmate's body. Overlapping the sheets of paper, use tape to secure the paper. Small body parts, such as fingers and toes, do not need to be wrapped individually. Cover entire hands and feet.

4. After your classmate is completely covered with paper, carefully cut the newspaper off his or her body. **WARNING:** *Do not cut any clothing or skin.*

5. Lay all of the overlapping sheets of newspaper on the floor. Using scissors and more tape, cut and piece the paper suit together to form a rectangle.

6. Using a meterstick, measure the length and width of the resulting rectangle. Multiply these two measurements for an estimate of the surface area of your classmate's skin.

Conclude and Apply

1. Was your estimation correct? Explain.

2. How accurate are your measurements of your classmate's skin surface area? How could your measurements be improved?

3. Calculate the volume of your classmate's skin, using 2 mm as the average thickness and your calculated surface area from this activity.

Communicating Your Data

Using a table, record the estimated skin surface area from all the groups in your class. Find the average surface areas for both males and females. Discuss any differences in these two averages. **For more help, refer to the Math Skill Handbook.**

Activity Use the Internet

Similar Skeletons

Humans and other mammals share many similar characteristics, including similar skeletal structures. Think about all the different types of mammals you have seen or read about. Tigers, dogs, and household cats are meat-eating mammals. Whales and dolphins live in water. Primates, which include gorillas, chimpanzees, and humans, walk on two legs. Mammals live in different environments, eat different types of food, and even look different, but they all have hair, possess the ability to maintain fairly constant body temperatures, and have similar skeletal structures.

Recognize the Problem

Which skeletal structures are similar among humans and other mammals? How many bones do you have in your hand? What types of bones are they? Do other mammals have similar skeletal structures?

Form a Hypothesis

Make a hypothesis about the skeletal structures that humans and other mammals have in common.

Goals
- **Identify** a skeletal structure in the human body.
- **Write** a list of mammals with which you are familiar.
- **Compare** the identified human skeletal structure to a skeletal structure in each of the mammals.
- **Determine** if the mammal skeletal structure that you selected is similar to the human skeletal structure you identified.

- **Describe** how the mammal skeletal structure is similar to or different from the skeletal structure in a human.

Data Source
SCIENCE *Online*
Visit the Glencoe Science Web site at **science.glencoe.com** to get more information about skeletal structures and for data collected by other students.

Test Your Hypothesis

Plan

1. Choose a specific part of the human skeletal structure to study, such as your hand, foot, skull, leg, or arm.
2. List four to six different mammals.
3. Do these mammals possess skeletal structures similar to the human skeleton? Remember, the mammals' skeletons can be similar to that of the human, but the structures can have different functions.
4. **Compare and contrast** the mammal and human skeletal structures. Are the types of bone similar? Is the number of bones the same? Where are these structures located?

Do

1. Make sure your teacher approves your plan before you start.
2. Go to the Glencoe Science Web site at **science.glencoe.com** to post your data.

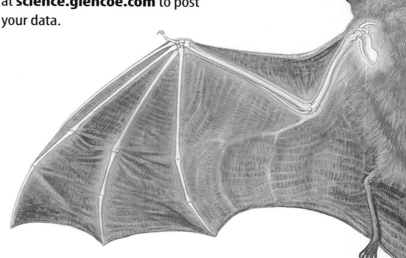

Analyze Your Data

1. Is each mammal's skeletal structure similar to or different from the human skeletal structure you chose?
2. In the data table provided on the Web site, describe how the structures are alike or different.

Draw Conclusions

1. Go to the Glencoe Science Web site at **science.glencoe.com** and compare your data to that of other students. Do other students agree with your conclusions?
2. Do the structures studied have similar function in the human and the mammals you researched?

Communicating
Your Data

SCIENCE *Online* Find this *Use the Internet* activity on the Glencoe Science Web site at **science.glencoe. com** and post your data in the table provided. **Compare** your data with that posted by other students.

First

A fashion doll is doing her part for medical science! It turns out that the plastic joints that make it possible for one type of doll's legs to bend make good joints in prosthetic (artificial) fingers in humans.

Jane Bahor (photo above) works at Duke University Medical Center in Durham, North Carolina. She makes lifelike body parts for people who have lost legs, arms, or fingers. A few years ago, she met a patient named Jennifer Jordan, an engineering student who'd lost a finger. The artificial finger that Bahor made looked real, but it couldn't bend. She and Jordan began to discuss the problem.

"If only the finger could bend, like a doll's legs bend," said Bahor. "It would be so much more useful to you!"

Jordan's eyes lit up. "That's it!" Jordan said. The engineer went home and borrowed one of her sister's dolls. Returning with it to Bahor's office, she and Bahor did "surgery." They operated on the fashion doll's legs and removed the knee joints from their vinyl casings.

"It turns out that the doll's knee joints flexed the same way that human finger joints do," says Bahor. "We could see that using these joints would allow patients more use and flexibility with their 'new' fingers."

Holding On

The new, fake, flexible fingers can bend in the same way that a doll's legs bend. A person can use his or her other hand to bend and straighten the joint. When the joint bends, it makes a sound similar to a cracking knuckle.

Being able to bend prosthetic fingers allows wearers to hold a pen, pick up a cup, or grab a steering wheel. These are tasks that were impossible before the plastic knee joints were implanted in the artificial fingers. "We've even figured out how to insert three joints in each finger, so that now its wearer can almost make a fist," adds Bahor. Just like the doll's legs, the prosthetic fingers stay bent until the wearer straightens them.

Bahor removes a knee joint from a doll. The joint will soon be in a human's finger!

Aid *Dolls*

Jane Bahor keeps a supply of dolls in her medicine cabinet. Some have had their knee joints removed.

A fashion doll helps improve the lives of people

Bahor called the company that makes the fashion doll and shared the surprising discovery. The toymaker was so impressed that Bahor now has a ten-year supply of plastic knee joints—free of charge!

But supplies come from other sources, too. "A Girl Scout troop in New Jersey just sent me a big box of donated dolls for the cause," reports Bahor. "It's really great to have kids' support in this effort."

CONNECTIONS **Invent** Choose a "problem" you can solve. Need a better place to store your notebooks in your locker, for instance? Use what Bahor calls "commonly found materials" to solve the problem. Then make a model or a drawing of the problem-solving device.

SCIENCE *Online*

For more information, visit
science.glencoe.com

Reviewing Main Ideas

Section 1 The Skeletal System

1. Bones are living structures that protect, support, make blood, store minerals, and provide for muscle attachment.

2. The skull and pelvic joints in adults do not move and are classified as immovable. *How does the shape of bones relate to their function?*

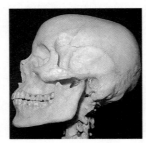

3. Movable joints move freely. Types of movable joints include pivot, hinge, ball-and-socket, and gliding joints.

Section 2 The Muscular System

1. Muscles contract to move bones and body parts.

2. Skeletal muscle is voluntary and moves bones. Smooth muscle is involuntary and controls movement of internal organs. Cardiac muscle is involuntary and located only in the heart.

3. Muscles contract—they pull, not push, to move body parts. Skeletal muscles work in pairs—when one contracts, the other relaxes. *How are skeletal muscles attached to the bones that they move?*

Section 3 The Skin

1. The epidermis has dead cells on its surface. Melanin is produced in the epidermis. Cells at the base of the epidermis produce new skin cells. The dermis is the inner layer where nerves, sweat and oil glands, and blood vessels are located.

2. The functions of skin include protection, reduction of water loss, production of vitamin D, and maintenance of body temperature. *What structures in the skin help maintain an even body temperature?*

3. Glands in the epidermis produce substances that destroy bacteria.

4. Severe damage to skin, including injuries and burns, can lead to infection and death if it is not treated. Advances in technology continue to provide different ways to repair such damage.

FOLDABLES
Reading & Study Skills

After You Read

Use the information in your Main Ideas Study Fold to review what you learned about skin, muscle, bone, and structure and movement from the chapter.

Visualizing Main Ideas

Complete the following concept map on body movement.

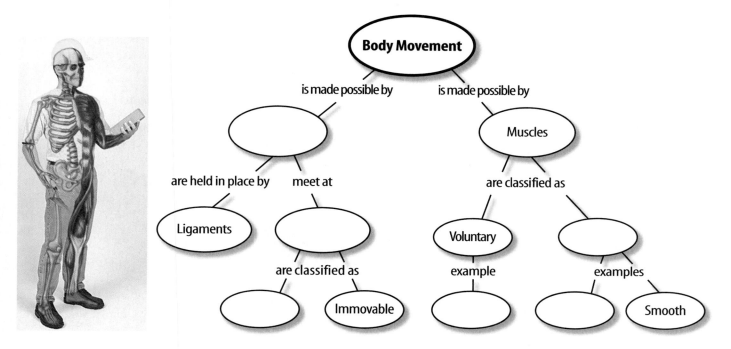

Vocabulary Review

Vocabulary Words

a. cardiac muscle
b. cartilage
c. dermis
d. epidermis
e. involuntary muscle
f. joint
g. ligament
h. melanin
i. muscle
j. periosteum
k. skeletal muscle
l. skeletal system
m. smooth muscle
n. tendon
o. voluntary muscle

THE PRINCETON REVIEW — Study Tip

Find a quiet place to study, whether at home or school. Turn off the television or radio, and give your full attention to your lessons.

Using Vocabulary

Match the definitions with the correct vocabulary word.

1. tough outer covering of bone

2. internal framework of the body

3. outer layer of skin

4. thick band of tissue that attaches muscle to a bone

5. muscle found in the heart

6. a tough band of tissue that holds two bones together

7. organ that can relax and contract to aid in the movement of the body

8. a muscle that you control

Checking Concepts

Choose the word or phrase that best answers the question.

1. Which of the following is the most solid form of bone?
   A) compact
   B) periosteum
   C) spongy
   D) marrow

2. Where are blood cells made?
   A) compact bone
   B) periosteum
   C) cartilage
   D) marrow

3. Where are minerals stored?
   A) bone
   B) skin
   C) muscle
   D) blood

4. What are the ends of bones covered with?
   A) cartilage
   B) tendons
   C) ligaments
   D) muscle

5. Where are immovable joints found in the human body?
   A) at the elbow
   B) at the neck
   C) in the wrist
   D) in the skull

6. What kind of joints are the knees, toes, and fingers?
   A) pivot
   B) hinge
   C) gliding
   D) ball and socket

7. Which vitamin is made in the skin?
   A) A
   B) B
   C) D
   D) K

8. Where are dead skin cells found?
   A) dermis
   B) marrow
   C) epidermis
   D) periosteum

9. Which of the following is found in bone?
   A) iron
   B) calcium
   C) vitamin D
   D) vitamin K

10. Which of the following structures helps retain fluids in the body?
    A) bone
    B) muscle
    C) skin
    D) a joint

Thinking Critically

11. When might skin not be able to produce enough vitamin D?

12. What factors might a doctor consider before choosing a method of skin repair for a severe burn victim?

13. What would lack of calcium do to bones?

14. How can you distinguish among these three muscle types?

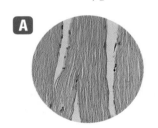 A

 B

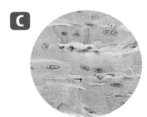 C

15. What function of your lower lip's skin changes when a dentist gives you novocaine before filling a bottom tooth? Why?

Developing Skills

16. **Drawing Conclusions** The joints in the skull of a newborn baby are flexible, but those of a teenager have fused together and are immovable. Conclude why the infant's skull joints are flexible.

17. **Predicting** Predict what would happen if a person's sweat glands didn't produce sweat.

18. **Comparing and Contrasting** Compare and contrast the functions of ligaments and tendons.

19. Forming Hypotheses Your body has about 3 million sweat glands. Make a hypothesis about where these sweat glands are on your body. Are they distributed evenly throughout your body?

20. Concept Mapping Complete the following concept map that describes the types and functions of bone cells.

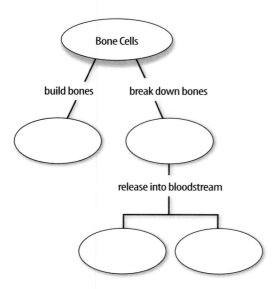

Performance Assessment

21. Display Research the differences among first-, second-, and third-degree burns. A local hospital's burn unit or fire department is a possible source of information about burns. Display pictures of each type of burn and descriptions of treatments on a three-sided, free-standing poster.

TECHNOLOGY

Go to the Glencoe Science Web site at **science.glencoe.com** or use the **Glencoe Science CD-ROM** for additional chapter assessment.

THE PRINCETON REVIEW **Test Practice**

Mia has researched the number of bones in different regions of the body. She placed her results in the following bar graph.

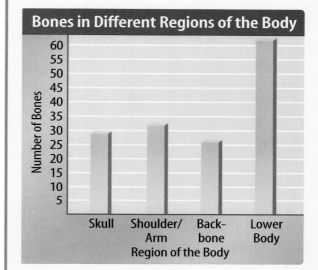

1. According to the graph, the region of the body that has the greatest number of bones is the _____.
A) backbone
B) skull
C) shoulder/arm
D) lower body

2. The total number of bones in the human body is 206. Approximately what percentage of bones is located in the backbone?
F) 2%
G) 12%
H) 50%
J) 75%

Nutrients and Digestion

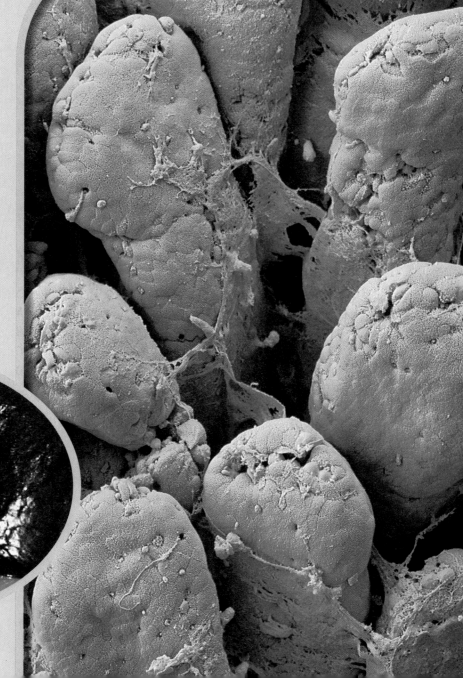

You may think the photograph on the right is a view of a pile of potatoes or a heap of loaves of bread, but it is a close-up of your small intestine. The wall of the small intestine has many fingerlike projections that absorb, or soak up, substances from digested food. Other organs in your body also help break down the food you eat.

What do you think?

Science Journal Look at this picture with a classmate. What part of the digestive system do you think it is? Here's a hint: *The fluid from this organ acts like soapy water does on a greasy pan.* Write your answer or best guess in your Science Journal.

EXPLORE **A**CTIVITY

Think about your favorite food. Now imagine yourself taking a bite of it. When you eat, your body breaks down food to release energy. How long does it take for the food to go through the entire process?

Model the digestive tract

1. Make a label for each of the major organs of the digestive tract listed here. Include the organ's name, its length, and the time it takes for food to pass through it.

2. Working with a partner, place a piece of masking tape that is 6.5 m long on the classroom floor.

3. Beginning at one end of the tape, and in the same order as they are listed in the table, mark the length for each organ. Place each label next to its section.

Organs of the Digestive System		
Organ	**Length**	**Time**
Mouth	8 cm	5 s to 30 s
Pharynx and Esophagus	25 cm	10 s
Stomach	16 cm	2 h to 4 h
Small Intestine	4.75 m	3 h
Large Intestine	1.25 m	2 days

Observe

In your Science Journal, suggest reasons why the food that you eat spends a different amount of time in each of the organs. What factors might change the amount of time digestion takes?

Before You Read

FOLDABLES
Reading &Study
Skills

Making a Classify Study Fold Make the following Foldable to help you organize foods based on the nutrients that they contain.

1. Place a sheet of paper in front of you so the short side is at the top. Fold the top of the paper down and the bottom up to divide the paper into thirds. Then fold the paper in half from top to bottom.

2. Open the paper and label the six columns as shown: *Proteins, Carbohydrates, Lipids, Water, Vitamins,* and *Minerals*.

3. As you read the chapter, list foods you eat that provide each of these nutrients in the proper columns.

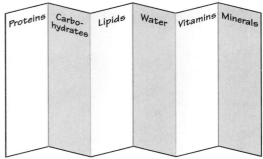

① Nutrition

As You Read

What You'll Learn

- **Distinguish** among the six classes of nutrients.
- **Identify** the importance of each type of nutrient.
- **Explain** the relationship between diet and health.

Vocabulary

nutrient   fat
protein   vitamin
amino acid   mineral
carbohydrate   food group

Why It's Important

You can make healthful food choices if you know what nutrients your body uses daily.

Figure 1
Foods vary in the number of Calories they contain. A hamburger has the same number of Calories as 8.5 average-sized carrots.

Why do you eat?

You're listening to a favorite song on the radio, maybe even singing along. Then all of a sudden, the music stops. You examine the radio to see what happened. The batteries died. You hunt for more batteries and quickly put in the new ones. In the same way that the radio needs batteries to work, you need food to carry out your daily activities—but not just any food. When you are hungry, you probably choose food based on taste and the amount of time you have to eat it. However, as much as you don't want to admit it, the nutritional value of the food you choose is more important than the taste. A chocolate-iced donut might be tasty and quick to eat, yet it provides few of the nutrients your body needs. **Nutrients** (NEW tree unts) are substances in foods that provide energy and materials for cell development, growth, and repair.

Energy Needs Your body needs energy for every activity that it performs. Muscle activities such as the beating of your heart, blinking your eyes, and lifting your backpack require energy. Your body also uses energy to maintain a steady internal temperature of about 37°C (98.6°F). This energy comes from the foods you eat. The amount of energy available in food is measured in Calories. A Calorie (Cal) is the amount of heat necessary to raise the temperature of 1 kg of water 1°C. As shown in **Figure 1,** different foods contain different numbers of Calories. A raw carrot may have 30 Cal. This means that when you eat a carrot, your body has 30 Cal of energy available to use. A slice of cheese pizza might have 170 Cal, and one hamburger might have 260 Cal. The number of Calories varies due to the kinds of nutrients a food provides.

Classes of Nutrients

Six kinds of nutrients are available in food—proteins, carbohydrates, fats, vitamins, minerals, and water. Proteins, carbohydrates, vitamins, and fats all contain carbon and are called organic nutrients. In contrast, inorganic nutrients, such as water and minerals, do not contain carbon. Foods containing carbohydrates, fats, and proteins need to be digested or broken down before your body can use them. Water, vitamins, and minerals don't require digestion and are absorbed directly into your bloodstream.

Figure 2
Meats, poultry, eggs, fish, peas, beans, and nuts are all rich in protein.

Proteins Your body uses proteins for replacement and repair of body cells and for growth. **Proteins** are large molecules that contain carbon, hydrogen, oxygen, nitrogen and sometimes sulfur. A molecule of protein is made up of a large number of smaller units, or building blocks, called **amino acids.** In **Figure 2** you can see some sources of proteins. Different foods contain different amounts of protein, as shown in **Figure 3.**

Your body needs only 20 amino acids in various combinations to make the thousands of proteins used in your cells. Most of these amino acids can be made in your body's cells, but eight of them cannot. These eight are called essential amino acids. They have to be supplied by the foods you eat. Complete proteins provide all of the essential amino acids. Eggs, milk, cheese, and meat contain complete proteins. Incomplete proteins are missing one or more of the essential amino acids. If you are a vegetarian, you can get all of the essential amino acids by eating a wide variety of protein-rich vegetables, fruits, and grains.

540 Calories
10 g protein

280 Calories
16 g protein

186 Calories
15 g protein

Figure 3
The amount of protein in a food is not the same as the number of Calories in the food. A taco has nearly the same amount of protein as a slice of pizza, but it usually has about 100 fewer Calories.

Figure 4
These foods contain carbohydrates that provide energy for all the things that you do.

Chemistry INTEGRATION

Carbohydrates Study the nutrition label on several boxes of cereal. You'll notice that the number of grams of carbohydrates found in a typical serving of cereal is higher than the amounts of the other nutrients. **Carbohydrates** (kar boh HI drayts) usually are the main sources of energy for your body. Each carbohydrate molecule is made of carbon, hydrogen, and oxygen atoms. Energy holds the atoms together. When carbohydrates are broken down in the presence of oxygen in your cells, this energy is released for use by your body.

Three types of carbohydrates are sugar, starch, and fiber, as shown in **Figure 4.** Sugars are called *simple carbohydrates.* You're probably most familiar with table sugar. However, fruits, honey, and milk also contain forms of sugar. Your cells break down glucose, a simple sugar. The other two types of carbohydrates— starch and fiber—are called *complex carbohydrates.* Starch is found in potatoes and foods made from grains such as pasta. Starches are made up of many simple sugars in long chains. Fiber, such as cellulose, is found in the cell walls of plant cells. Foods like whole-grain breads and cereals, beans, peas, and other vegetables and fruits are good sources of fiber. Because different types of fiber are found in foods, you should eat a variety of fiber-rich plant foods. You cannot digest fiber, but it is needed to keep your digestive system running smoothly.

Nutritious snacks can help your body get the nutrients it needs, especially when you are growing rapidly and are physically active. Choose snacks that provide nutrients such as complex carbohydrates, proteins, and vitamins, as well as fiber. Foods high in sugar and fat can have lots of Calories that supply energy, but they provide only some of the nutrients your body needs.

Figure 5

Fat is stored in certain cells in your body. **A** The cyto-plasm and nucleus are pushed to the edge of the cell by the fat deposits. **B** Some foods you might choose for lunch or snacks that are high in fat are outlined in red.

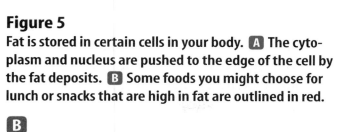

Fats The term fat has developed a negative meaning for some people. However, **fats,** also called lipids, are necessary because they provide energy and help your body absorb vitamins. Fat tissue cushions your internal organs. A major part of every cell membrane is made up of fat. A gram of fat can release more than twice as much energy as a gram of carbohydrate can. During the digestion process, fat is broken down into smaller molecules called fatty acids and glycerol (GLIHS ur awl). Because fat is a good storage unit for energy, excess energy from the foods you eat is converted to fat and stored for later use, as shown in **Figure 5A.**

> ✔ **Reading Check** *Why is fat a good storage unit for energy?*

Fats are classified as unsaturated or saturated based on their chemical structure. Unsaturated fats are usually liquid at room temperature. Vegetable oils as well as fats found in seeds are unsaturated fats. Saturated fats are found in meats, animal prod-ucts, and some plants and are usually solid at room temperature. Although fish contains saturated fat, it also has some unsaturated fats that your body needs. Saturated fats have been associated with high levels of blood cholesterol. Your body makes choles-terol in your liver. Cholesterol is part of the cell membrane in all of your cells. However, a diet high in cholesterol may result in deposits forming on the inside walls of blood vessels. These deposits can block the blood supply to organs and increase blood pressure. This can lead to heart disease and strokes.

Mini LAB

Comparing the Fat Content of Foods

Procedure 🥽 👕 🚫

1. Collect three pieces of each of the following foods: **potato chips; pretzels; peanuts;** and **small cubes of fruits, cheese, vegeta-bles, and meat.**
2. Place the food items on a piece of **brown grocery bag.** Label the paper with the name of each food. Do not taste the foods.
3. Allow foods to sit for 30 min.
4. Remove the items, properly dispose of them, and observe the paper.

Analysis

1. Which items left a translucent (greasy) mark? Which left a wet spot?
2. How are the foods that left a greasy spot on the paper alike?
3. Use this test to determine which other foods contain fats. A greasy mark means the food contains fat. A wet mark means the food contains a lot of water.

Vitamins Those organic nutrients needed in small quantities for growth, regulating body functions, and preventing some diseases are called **vitamins.** For instance, your bone cells need vitamin D to use calcium, and your blood needs vitamin K in order to clot.

Most foods supply some vitamins, but no food has them all. Some people feel that taking extra vitamins is helpful, while others feel that eating a well-balanced diet usually gives your body all the vitamins it needs.

Vitamins are classified into two groups, as shown in **Figure 6.** Some vitamins dissolve easily in water and are called water-soluble vitamins. They are not stored by your body so you have to take them daily. Other vitamins dissolve only in fat and are called fat-soluble vitamins. These vitamins are stored by your body. Although you eat or drink most vitamins, some are made by your body. Vitamin D is made when your skin is exposed to sunlight. Some vitamin K and two of the B vitamins are made with the help of bacteria that live in your large intestine.

Problem-Solving Activity

Is it unhealthy to snack between meals?

Most children eat three meals each day accompanied by snacks in between. Grabbing a bite to eat to satisfy you until your next meal is a common occurrence in today's society, and 20% of our energy and nutrient needs comes from snacking. While it would be best to select snacks consisting of fruits and vegetables, most children prefer to eat a bag of chips or a candy bar. Although these quick snacks are highly convenient, many times they are high in fat, as well.

Identifying the Problem

The table on the right lists several snack foods that are popular among adolescents. They are listed alphabetically, and the grams of fat per individual serving is shown. As you examine the chart, can you conclude which snacks would be a healthier choice based on their fat content?

Fat in Snack Foods	
One Serving	**Fat (g)**
Candy bar	12
Frozen pizza	30
Ice cream	8
Potato chips	10
Pretzels	1

1. Looking at the data, what can you conclude about the snack foods you eat? What other snack foods do you eat that are not listed on the chart? How do you think they compare in nutritional value? Which snack foods are healthiest?

2. Pizza appears to be the unhealthiest choice on the chart because of the amount of the fat it contains. Why do you think pizza contains so much fat? List at least three ways to make pizza a healthier snack food.

Figure 6

Vitamins come in two groups—water-soluble, which should be replaced daily, and fat-soluble, which can be stored in the body. The sources and benefits of both groups are shown below.

WATER-SOLUBLE
Need to be replenished every day because they are excreted by the body

Aids in growth, healthy nervous system, use of carbohydrates, and red blood cell production

B

(B₆, B₁₂, riboflavin, niacin, thiamine, etc.)

Aids in growth, healthy bones and teeth, wound recovery

C

Aids in growth, eyesight, healthy skin

FAT-SOLUBLE
Stored in the body in fatty tissue

Aids in absorption of calcium and phosphorus by bones and teeth

A

D

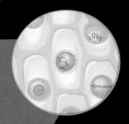

Aids in formation of cell membranes

Aids in blood clotting and wound recovery

E

K

Minerals Inorganic nutrients—nutrients that lack carbon and regulate many chemical reactions in your body—are called **minerals.** Your body uses about 14 minerals. Minerals build cells, take part in chemical reactions in cells, send nerve impulses throughout your body, and carry oxygen to body cells. In **Figure 7,** you can see how minerals can get from the soil into your body. Of the 14 minerals, calcium and phosphorus are used in the largest amounts for a variety of body functions. One of these functions is the formation and maintenance of bone. Some minerals, called trace minerals, are required only in small amounts. Copper and iodine usually are listed as trace minerals. Several minerals, what they do, and some food sources for them are listed in **Table 1.**

Figure 7
The roots of the wheat take in phosphorus from the soil. Then the mature wheat is harvested and used in bread and cereal. Your body gets phosphorus when you eat the cereal.

✔ **Reading Check** *Why is copper considered a trace mineral?*

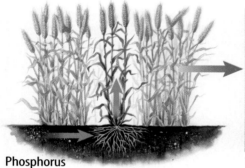
Phosphorus

Wheat being harvested

Table 1 Minerals

Mineral	Health Effect	Food Sources
Calcium	strong bones and teeth, blood clotting, muscle and nerve activity	dairy products, eggs, green leafy vegetables, soy
Phosphorus	strong bones and teeth, muscle contraction, stores energy	cheese, meat, cereal
Potassium	balance of water in cells, nerve impulse conduction, muscle contraction	bananas, potatoes, nuts, meat, oranges
Sodium	fluid balance in tissues, nerve impulse conduction	meat, milk, cheese, salt, beets, carrots, nearly all foods
Iron	oxygen is transported in hemoglobin by red blood cells	red meat, raisins, beans, spinach, eggs
Iodine (trace)	thyroid activity, metabolic stimulation	seafood, iodized salt

Figure 8
About two thirds of your body water is located within your body cells. Water helps maintain the cells' shapes and sizes. The water that is lost through perspiration and respiration must be replaced.

Water Loss	
Method of Loss	**Amount (mL/day)**
Exhaled air	350
Feces	150
Skin (mostly as sweat)	500
Urine	1,500

Water Have you ever gone on a bike ride on a hot summer day without a bottle of water? You probably were thirsty and maybe you even stopped to get some water. Water is important for your body. Next to oxygen, water is the most important factor for survival. Different organisms need different amounts of water to survive. You could live for a few weeks without food but for only a few days without water because your cells need water to carry out their work. Most of the nutrients you have studied in this chapter can't be used by your body unless they are carried in a solution. This means that they have to be dissolved in water. In cells, chemical reactions take place in solutions.

The human body is about 60 percent water by weight. About two thirds of your body water is located in your body cells. Water also is found around cells and in body fluids such as blood. As shown in **Figure 8,** your body loses water as perspiration. When you exhale, water leaves your body as water vapor. Water also is lost every day when your body gets rid of wastes. To replace water lost each day, you need to drink about 2 L of liquids. However, drinking liquids isn't the only way to supply cells with water. Most foods have more water than you realize. An apple is about 80 percent water, and many meats are 90 percent water.

Earth Science
INTEGRATION

The mineral halite is processed to make table salt. In the United States, most salt comes from underground mines. Research to find the locations of these mines, then label them on a map.

Why do you get thirsty? Your body is made up of systems that operate together. When your body needs to replace lost water, messages are sent to your brain that result in a feeling of thirst. Drinking water satisfies your thirst and usually restores the body's homeostasis (hoh mee oh STAY sus). Homeostasis is the regulation of the body's internal environment, such as temperature and amount of water. When homeostasis is restored, the signal to the brain stops and you no longer feel thirsty.

Food Groups

Because no naturally occurring food has every nutrient, you need to eat a variety of foods. Nutritionists have developed a simple system, called the food pyramid, shown in **Figure 9,** to help people select foods that supply all the nutrients needed for energy and growth.

Foods that contain the same type of nutrient belong to a **food group.** Foods have been divided into five groups—bread and cereal, vegetable, fruit, milk, and meat. The recommended daily amount for each food group will supply your body with the nutrients it needs for good health. Using the food pyramid to make choices when you eat will help you maintain good health.

Figure 9
The pyramid shape reminds you that you should consume more servings from the bread and cereal group than from other groups.
Where should the least number of servings come from?

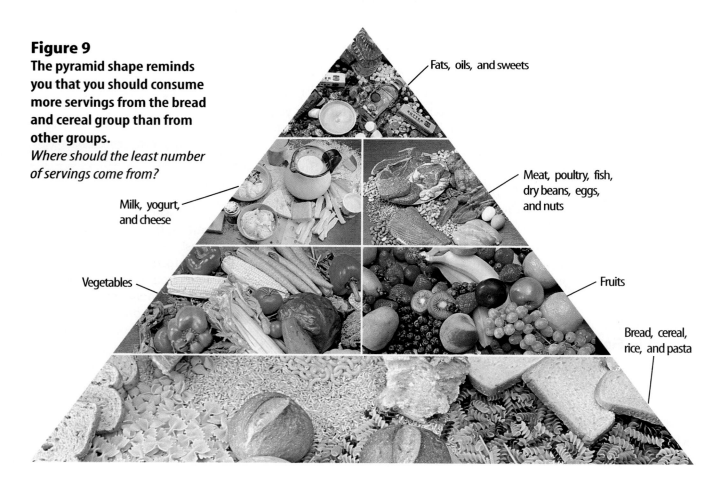

Fats, oils, and sweets

Meat, poultry, fish, dry beans, eggs, and nuts

Milk, yogurt, and cheese

Vegetables

Fruits

Bread, cereal, rice, and pasta

Daily Servings Each day you should eat six to eleven servings from the bread and cereal group, three to five servings from the vegetable group, two to four servings from the fruit group, two to three servings from the milk group, and two to three servings from the meat and beans group. Only small amounts of fats, oils, and sweets should be consumed.

The size of a serving is different for different foods. For example, a slice of bread or one ounce of ready-to-eat cereal is a bread-and-cereal group serving. One cup of raw leafy vegetables or one-half cup of cooked or chopped raw vegetables make a serving from the vegetable group. One medium apple, banana, or orange is a fruit serving. A serving from the milk group can be one cup of milk or yogurt. Two ounces of cooked lean meat, one-half cup of cooked dry beans, or one egg counts as a serving from the meat and beans group.

Food Labels The nutritional facts found on all packaged foods make it easier to make healthful food choices. These labels, as shown in **Figure 10,** can help you plan meals that supply the daily recommended amounts of nutrients and meet special dietary requirements (for example, a low-fat diet).

Figure 10
The information on a food label can help you decide what to eat.

Nutrition Facts
Serving Size 1 Meal

Amount Per Serving

Calories 330   Calories from Fat 60

	% Daily Value*
Total Fat 7g	**10%**
Saturated Fat 3.5g	**17%**
Polyunsaturated Fat 1g	
Monounsaturated Fat 2.5g	
Cholesterol 35mg	**12%**
Sodium 460mg	**19%**
Total Carbohydrate 52g	**18%**
Dietary Fiber 6g	**24%**
Sugars 17g	
Protein 15g	

Vitamin A 15%	Vitamin C 70%
Calcium 4%	Iron 10%

* Percent Daily Values are based on a 2,000 calorie diet. Your daily values may be higher or lower depending on your calorie needs.

	Calories	2,000	2,500
Total Fat	Less than	65g	80g
Sat Fat	Less than	20g	25g
Cholesterol	Less than	300mg	300mg
Sodium	Less than	2,400mg	2,400mg
Total Carbohydrate		300g	375g
Dietary Fiber		25g	30g

Section ① Assessment

1. List six classes of nutrients that your body needs and give one example of a food source for each.

2. Describe a major function of each class of nutrient.

3. Discuss how food choices can positively and negatively affect your health.

4. Explain the importance of water in the body.

5. **Think Critically** What foods from each food group would provide a balanced breakfast? Explain.

Skill Builder Activities

6. **Interpreting Data** Nutritional information can be found on the labels of most foods. Interpret the labels found on three different types of food products. **For more help, refer to the** Science Skill Handbook.

7. **Using an Electronic Spreadsheet** Make a spreadsheet of the minerals listed in **Table 1.** Use reference books to gather information about minerals and add these to the table: *sulfur, magnesium, copper, manganese, cobalt,* and *zinc.* **For more help, refer to the** Technology Skill Handbook.

Activity

Identifying Vitamin C Content

Vitamin C is found in many fruits and vegetables. Oranges have a high vitamin C content. Try this activity to test which orange juice has the highest vitamin C content.

What You'll Investigate
Which orange juice contains the most vitamin C?

Materials
test tube (4)	2% tincture of iodine
*paper cups	dropper
test-tube rack	cornstarch
masking tape	triple-beam balance
wooden stirrer (13)	weighing paper
graduated cylinder	water (50 mL)
*graduated container	glass-marking pencil

dropper bottles (4) containing:
    (1) freshly squeezed orange juice
    (2) orange juice made from frozen concentrate
    (3) canned orange juice
    (4) dairy carton orange juice

* Alternate materials

Goals
■ **Observe** the vitamin C content of different orange juices.

Safety Precautions

Wear your goggles and apron. Do not taste any of the juices. Iodine is poisonous and can stain skin and clothing. It is an irritant and can cause damage if it comes in contact with your eyes. Notify your teacher if a spill occurs.

Procedure

1. Make a data table like the example shown to record your observations.

Drops of Iodine Needed to Change Color				
Juice	**Trial**			**Average**
	1	**2**	**3**	
1 Fresh Juice				
2 Frozen Juice				
3 Canned Juice				
4 Carton Juice				

2. Label four test tubes 1 through 4 and place them in the test-tube rack.

3. **Measure** and pour 5 mL of juice from bottle 1 into test tube 1, 5 mL from bottle 2 into test tube 2, 5 mL from bottle 3 into test tube 3, and 5 mL from bottle 4 into test tube 4.

4. Weigh 0.3 g of cornstarch, then put it in a container. Slowly mix in 50 mL of water until the cornstarch completely dissolves.

5. Add 5 mL of the cornstarch solution to each of the four test tubes. Stir well.

6. Add iodine to test tube 1, one drop at a time. Stir after each drop. Record the number of drops it takes for the juice to change to a purple color. The more vitamin C that is present, the more drops it takes to change color.

7. Repeat step 6 with test tubes 2, 3, and 4.

8. Empty and clean the test tubes. Repeat steps 3 through 7 two more times, then average your results.

Conclude and Apply

1. **Compare and contrast** the amount of vitamin C in the orange juices tested.

2. If the amount of vitamin C varied in the orange juices, suggest a reason why. Check

SECTION 2
The Digestive System

Functions of the Digestive System

You are walking through a park on a cool, autumn afternoon. Birds are searching in the grass for insects. A squirrel is eating an acorn. Why are the animals so busy? Like you, they need food to supply their bodies with energy. Food is processed in your body in four stages—ingestion, digestion, absorption, and elimination. Whether it is a piece of fruit or an entire meal, all the food you eat is treated to the same processes in your body. As soon as food enters your mouth, or is ingested as shown in **Figure 11,** breakdown begins. **Digestion** is the process that breaks down food into small molecules so that they can be absorbed and moved into the blood. From the blood, food molecules are transported across the cell membrane to be used by the cell. Unused molecules pass out of your body as wastes.

Digestion is mechanical and chemical. **Mechanical digestion** takes place when food is chewed, mixed, and churned. **Chemical digestion** occurs when chemical reactions occur that break down large molecules of food into smaller ones.

As You Read

What You'll Learn
- **Distinguish** the differences between mechanical digestion and chemical digestion.
- **Identify** the organs of the digestive system and what takes place in each.
- **Explain** how homeostasis is maintained in digestion.

Vocabulary
digestion	peristalsis
mechanical digestion	chyme
chemical digestion	villi
enzyme	

Why It's Important
The processes of the digestive system make the food you eat available to your cells.

Figure 11
Humans have to chew solid foods before swallowing them, but snakes have adaptations that allow them to swallow their food whole.

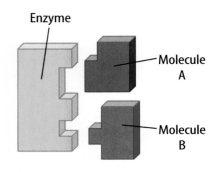

A The surface shape of an enzyme fits the shape of specific molecules that take part in the reaction.

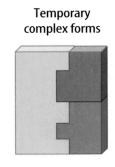

B The enzyme and the molecules join and the reaction occurs between the two molecules.

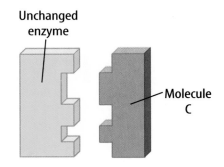

C Following the reaction, the enzyme and the new molecule separate. The enzyme is not changed by the reaction. The resulting new molecule has a new chemical structure.

Figure 12
Enzymes speed up the rate of certain body reactions. During these reactions, the enzymes are not used up or changed in any way. *What happens to the enzyme after it separates from the new molecule?*

Enzymes

Chemical digestion is possible only because of enzymes (EN zimez). An **enzyme** is a type of protein that speeds up the rate of a chemical reaction in your body. One way enzymes speed up reactions is by reducing the amount of energy necessary for a chemical reaction to begin. If enzymes weren't there to help, the rate of chemical reactions would slow down. Some might not even happen at all. As shown in **Figure 12,** enzymes work without being changed or used up.

Enzymes in Digestion Many enzymes help you digest carbohydrates, proteins, and fats. Amylase (AM uh lays) is an enzyme produced by glands near the mouth. This enzyme helps speed up the breakdown of complex carbohydrates, such as starch, into simpler carbohydrates—sugars. In your stomach, the enzyme pepsin aids the chemical reactions that break down complex proteins into less complex proteins. In your small intestine, a number of other enzymes continue to speed up the breakdown of proteins into amino acids. The pancreas, an organ on the back side of the stomach, releases several enzymes through a tube into the small intestine. Some of these enzymes continue to aid the process of starch breakdown that started in the mouth. The resulting sugars are turned into glucose and are used by your body's cells. Different enzymes from the pancreas are involved in the breakdown of fats into fatty acids. Others help in the reactions that break down proteins.

 **Reading Check** *What is the role of enzymes in the chemical digestion of food?*

Other Enzyme Actions Enzyme-aided reactions are not limited to the digestive process. Enzymes also help speed up chemical reactions responsible for building your body. They are involved in the energy production activities of your muscle and nerve cells. Enzymes also aid in the blood-clotting process. Without enzymes, the chemical reactions of your body would not happen. In fact, you would not exist.

Organs of the Digestive System

Your digestive system has two parts—the digestive tract and the accessory organs. The major organs of your digestive tract—mouth, esophagus (ih SAH fuh guhs), stomach, small intestine, large intestine, rectum, and anus—are shown in **Figure 13.** Food passes through all of these organs. The tongue, teeth, salivary glands, liver, gallbladder, and pancreas, also shown in **Figure 13,** are the accessory organs. Although food doesn't pass through them, they are important in mechanical and chemical digestion. Your liver, gallbladder, and pancreas produce or store enzymes and chemicals that help break down food as it passes through the digestive tract.

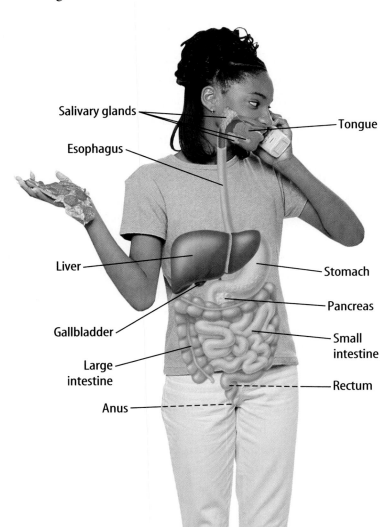

Salivary glands
Esophagus
Liver
Gallbladder
Large intestine
Anus
Tongue
Stomach
Pancreas
Small intestine
Rectum

Figure 13
The human digestive system can be described as a tube divided into several specialized sections. If stretched out, an adult's digestive system is 6 m to 9 m long.

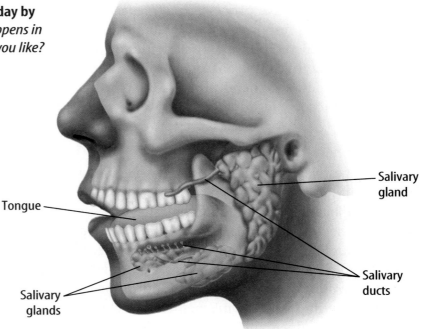

Figure 14
About 1.5 L of saliva are produced each day by salivary glands in your mouth. *What happens in your mouth when you think about a food you like?*

Tongue

Salivary gland

Salivary ducts

Salivary glands

Research Visit the Glencoe Science Web site at **science.glencoe.com** for more information about the role of the stomach during digestion. Communicate to your class what you learned.

The Mouth Mechanical and chemical digestion begin in your mouth. Mechanical digestion happens when you chew your food with your teeth and mix it with your tongue. Chemical digestion begins with the addition of a watery substance called saliva (suh LIVE uh). As you chew, your tongue moves food around and mixes it with saliva. Saliva is produced by three sets of glands near your mouth, as shown in **Figure 14.** Although saliva is mostly water, it also contains mucus and an enzyme that aids in the breakdown of starch into sugar. Food mixed with saliva becomes a soft mass and is moved to the back of your mouth by your tongue. It is swallowed and passes into your esophagus. Now ingestion is complete, but the process of digestion continues.

The Esophagus Food moving into the esophagus passes over the epiglottis (ep uh GLAHT us). This structure automatically covers the opening to the windpipe to prevent food from entering it, otherwise you would choke. Your esophagus is a muscular tube about 25 cm long. It takes about 4 s to 10 s for food to move down the esophagus to the stomach. No digestion takes place in the esophagus. Mucous glands in the wall of the esophagus keep the food moist. Smooth muscles in the wall move food downward with a squeezing action. These waves of muscle contractions, called **peristalsis** (per uh STAHL sus), move food through the entire digestive tract.

The Stomach The stomach, shown in **Figure 15,** is a muscular bag. When empty, it is somewhat sausage shaped with folds on the inside. As food enters from the esophagus, the stomach expands and the folds smooth out. Mechanical and chemical digestion take place in the stomach. Mechanically, food is mixed in the stomach by peristalsis. Chemically, food is mixed with enzymes and strong digestive solutions, such as hydrochloric acid, to help break it down.

Specialized cells in the walls of the stomach release about 2 L of hydrochloric acid each day. The acid works with the enzyme pepsin to digest protein. Hydrochloric acid has another important purpose—it destroys bacteria that are present in the food. The stomach also produces mucus, which makes food more slippery and protects the stomach from the strong, digestive solutions. Food moves through your stomach in 2 hours to 4 hours and is changed into a thin, watery liquid called **chyme** (KIME). Little by little, chyme moves out of your stomach and into your small intestine.

Reading Check *Why isn't your stomach digested by hydrochloric acid?*

Figure 15
A band of muscle is at the entrance of the stomach to control the entry of food from the esophagus. Muscles at the end of the stomach control the flow of the partially digested food into the first part of the small intestine.

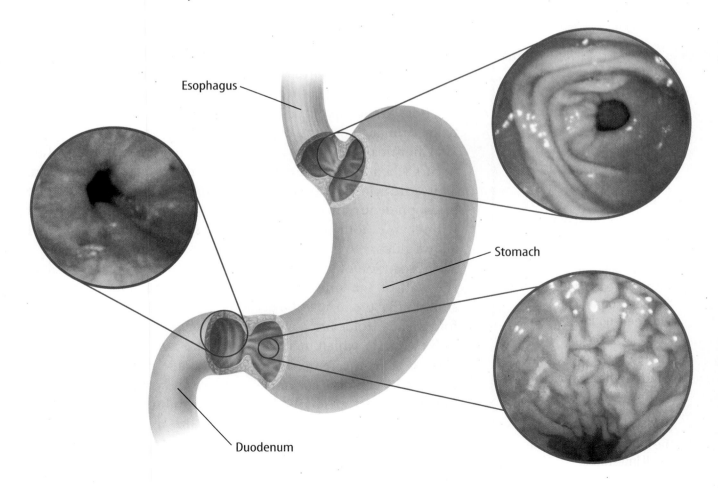

Esophagus

Stomach

Duodenum

Figure 16

Hundreds of thousands of densely packed villi give the impression of a velvet cloth surface. If the surface area of your villi could be stretched out, it would cover an area the size of a baseball diamond. *What would happen to a person's weight if the number of villi were drastically reduced? Why?*

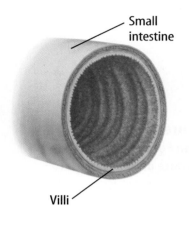

Small intestine

Villi

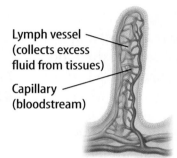

Lymph vessel (collects excess fluid from tissues)

Capillary (bloodstream)

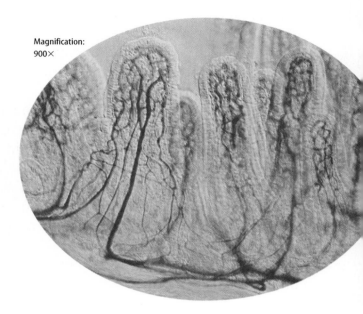

Magnification: 900×

TRY AT HOME Mini LAB

Modeling Absorption in the Small Intestine

Procedure 🖐️✋

1. Place one piece of **smooth cotton cloth** (about 25 cm × 25 cm) and a similar-sized piece of **cotton terry cloth** into a **bowl of water.**
2. Soak each for 30 s.
3. Remove the cloths and drain for 1 minute.
4. Wring out each cloth into different containers. Measure the amount of water collected in each.

Analysis

1. Which cloth absorbed the most water?
2. How does the surface of the terry cloth compare to the internal surface of the small intestine?

The Small Intestine Your small intestine is small in diameter, but it measures 4 m to 7 m in length. As chyme leaves your stomach, it enters the first part of your small intestine, called the duodenum (doo AHD un um). Most digestion takes place in your duodenum. Here, a greenish fluid from the liver, called bile, is added. The acid from the stomach makes large fat particles float to the top of the liquid. Bile breaks up the large fat particles, similar to the way detergent breaks up grease.

Chemical digestion of carbohydrates, proteins, and fats occurs when a digestive solution from the pancreas is mixed in. This solution contains bicarbonate ions and enzymes. The bicarbonate ions help neutralize the stomach acid that is mixed with chyme. Your pancreas also makes insulin, a hormone that allows glucose to pass from the bloodstream into your cells.

Absorption of food takes place in the small intestine. Look at the wall of the small intestine in **Figure 16.** The wall is not smooth like the inside of a garden hose but has many ridges and folds. These folds are covered with fingerlike projections called **villi** (VIHL i), which increase the surface area of the small intestine so that nutrients in the chyme have more places to be absorbed. Peristalsis continues to move and mix the chyme. The villi move and are bathed in the soupy liquid. Molecules of nutrients move into blood vessels within the villi. From here, blood transports the nutrients to all cells of your body. Peristalsis continues to force the remaining undigested and unabsorbed materials slowly into the large intestine.

The Large Intestine When the chyme enters the large intestine, it is still a thin, watery mixture. The main job of the large intestine is to absorb water from the undigested mass. This keeps large amounts of water in your body and helps maintain homeostasis. Peristalsis usually slows down in the large intestine. The chyme might stay there for as long as three days. After the excess water is absorbed, the remaining undigested materials become more solid. Muscles in the rectum, which is the last section of the large intestine, and the anus control the release of semisolid wastes from the body in the form of feces (FEE seez).

 **Reading Check** *Why does chyme remain in the large intestine for up to three days?*

Bacteria Are Important

Many types of bacteria live in your body. Bacteria live in many of the organs of your digestive tract including your mouth and large intestine. Some of these bacteria live in a relationship that is beneficial to the bacteria and to your body. The bacteria in your large intestine feed on undigested material like cellulose. In turn, bacteria make vitamins you need—vitamin K and two B vitamins. Vitamin K is needed for blood clotting. The two B vitamins, niacin and thiamine, are important for your nervous system and for other body functions. Bacterial action also converts bile pigments into new compounds. The breakdown of intestinal materials by bacteria produces gas.

 Environmental Science INTEGRATION

The species of bacteria that live in your large intestine are adapted to their habitat. What do you think would happen to the bacteria if their environment were to change? How would this affect your large intestine? Discuss your ideas with a classmate and write your answers in your Science Journal.

Section Assessment

1. Compare mechanical digestion and chemical digestion.
2. Name, in order, the organs through which food passes as it moves through the digestive tract.
3. How do activities in the large intestine help maintain homeostasis?
4. How do the accessory organs aid digestion?
5. **Think Critically** Crackers contain starch. Explain why a cracker begins to taste sweet after it is in your mouth for five minutes without being chewed.

Skill Builder Activities

6. **Recognizing Cause and Effect** What would happen to some of the nutrients in chyme if the pancreas did not secrete its solution into the small intestine? **For more help, refer to the** Science Skill Handbook.

7. **Communicating** Write a paragraph in your Science Journal explaining what would happen to the mechanical and chemical digestion in a person missing a large portion of his or her stomach. **For more help, refer to the** Science Skill Handbook.

Activity

Particle Size and Absorption

Before food reaches the small intestine, it is digested mechanically in the mouth and the stomach. The food mass is reduced to small particles. You can chew an apple into small pieces, but you would feed applesauce to a small child who didn't have teeth. What is the advantage of reducing the size of the food material?

What You'll Investigate

How does reducing the size of food particles aid the process of digestion?

Materials

beakers or jars (3)
thermometers (3)
sugar granules
mortar and pestle
triple beam balance
stirring rod
sugar cubes
weighing paper
warm water
stopwatch

Safety Precautions

Do not taste, eat, or drink any materials used in the lab.

Goals

- **Compare** the dissolving rates of different sized particles.
- **Predict** the dissolving rate of sugar particles larger than sugar cubes.
- **Predict** the dissolving rate of sugar particles smaller than particles of ground sugar.
- Using the lab results, **infer** why the body must break down and dissolve food particles.

Procedure

1. Copy the data table below into your Science Journal.

Dissolving Time of Sugar Particles		
Size of Sugar Particles	Mass	Time Until Dissolved
Sugar cube		
Sugar granules		
Ground sugar particles		

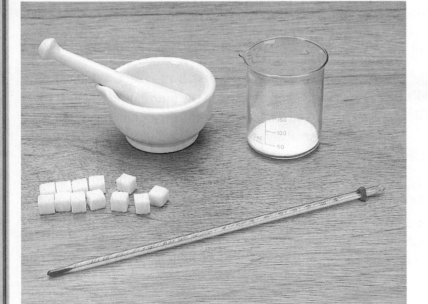

Procedure

2. Place a sugar cube into your mortar and grind up the cube with the pestle until the sugar becomes powder.

3. Using the triple-beam balance and weighing paper, measure the mass of the powdered sugar from your mortar. Using separate sheets of weighing paper, measure the mass of a sugar cube and the mass of a sample of the granular sugar. The masses of the powdered sugar, sugar cube, and granular sugar should be approximately equal to each other. Record the three masses in your data table.

4. Place warm water into the three beakers. Use the thermometers to be certain the water in each beaker is the same temperature.

5. Place the sugar cube in a beaker, the powdered sugar in a second beaker, and the granular sugar in the third beaker. Place all the sugar samples in the beakers at the same time and start the stopwatch when you put the sugar samples in the beaker.

6. Stir each sample equally.

7. **Measure** the time it takes each sugar sample to dissolve and record the times in your data table.

Conclude and Apply

1. **Identify** the experiment's constants and variables.

2. **Compare** the rate at which the sugar samples dissolved. What type of sugar dissolved most rapidly? Which was the slowest to dissolve?

3. **Predict** how long it would take sugar particles larger than the sugar cubes to dissolve. Predict how long it would take sugar particles smaller than the powdered sugar to dissolve.

4. **Infer and explain** the reason why small particles dissolve more rapidly than large particles.

5. **Infer** why you should thoroughly chew your food.

6. **Explain** how reducing the size of food particles aids the process of digestion.

*C*ommunicating
Your Data

Write a news column for a health magazine explaining to health-conscious people what they can do to digest their food better.

Eating Well

Does the same diet work for everyone?

R. Rajalakshmi (RAH jah lok shmee) grew up in India in the first half of the twentieth century, seeing many people around her who did not get enough food. Breakfast for a poor child might have been a cup of tea. Lunch might have consisted of a slice of bread. For dinner, a child might have eaten a serving of rice with a small piece of fish. This type of diet, low in calories and nutrients, produced pencil-thin children who were often sick and died young.

Good Diet, Wrong Place

R. Rajalakshmi studied biochemistry and nutrition at universities in India and in Canada. In the 1960s, she was asked to help manage a program to improve nutrition in her country. At that time, most advice on nutrition came from North American and European countries. Nutritionists suggested foods that were common and worked well for people who lived in these nations.

For example, they told poor Indian women to eat more meat and eggs and drink more orange juice. But Rajalakshmi knew this advice was useless in a country such as India. People there didn't eat such foods. They weren't easy to find. And for the poor, such foods were too expensive.

The Proper Diet for India

Rajalakshmi knew that for the program to work, it had to fit Indian culture. So she decided to restructure the nutrition program. She first found out what healthy middle class people in India ate. She took note of the nutrients available in those foods. Then she looked for cheap, easy-to-find foods that would provide the same nutrients.

Rajalakshmi created a balanced diet of locally grown fruits, vegetables, and grains.

These foods were cheap and could be cooked with simple equipment. Legumes (plants related to peas and peanuts), vegetables, and an Indian food called dhokla (DOH kluh) were basics. Dhokla is made of grains, legumes, and leafy vegetables. The grains and legumes provided protein, and the vegetables added vitamins and minerals.

Rajalakshmi's ideas were thought unusual in the 1960s. For example, she insisted that a diet without meat could provide all major nutrients. Now we know she was right. But it took persistence to get others to accept her diet about 40 years ago. Because of Rajalakshmi's program, Indian children almost doubled their food intake. And many children who would have been hungry and ill grew healthy and strong.

Thanks to R. Rajalakshmi and other nutritionists, many children in India are eating well and staying healthy.

CONNECTIONS Report Choose a continent and research what foods are native to that area. Share your findings with your classmates and compile a list of the foods and where they originated. Using the class list, create a world map on a bulletin board that shows the origins of the different foods.

SCIENCE *Online*
For more information, visit
science.glencoe.com

Reviewing Main Ideas

Section 1 Nutrition

1. Proteins, carbohydrates, fats, vitamins, minerals, and water are the six nutrients found in foods.

2. Carbohydrates provide energy, proteins are needed for growth and repair, and fats store energy and cushion organs. Vitamins and minerals regulate functions. Water makes up about 60 percent of your body's mass and is used for a variety of homeostatic functions.

3. Health is affected by the combination of foods that make up a diet. *In the photograph below, which foods would you choose to get the highest amount of protein in your diet?*

Section 2 The Digestive System

1. Mechanical digestion breaks down food through chewing and churning. Enzymes and other chemicals aid chemical digestion. Both types of digestion are used to break down food into substances that cells can absorb and use. Carbohydrates become simple sugars, proteins become amino acids, and fats are digested into fatty acids and glycerol.

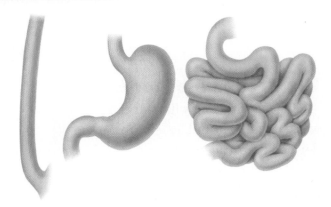

2. Food passes through the mouth, esophagus, stomach, small intestine, and large intestine and then out the anus. Ingestion takes place in the mouth. Digestion occurs in the mouth, stomach, and small intestine, with absorption occurring in the small and large intestines. Wastes are excreted through the anus. *In the illustration of the digestive organs shown above, where does digestion take place?*

3. The accessory digestive organs move and cut up food and supply digestive enzymes and other chemicals, such as bile, needed for digestion.

4. The large intestine absorbs water, which helps the body maintain homeostasis. Homeostasis is the regulation of the body's internal environment.

FOLDABLES
Reading & Study Skills

After You Read

Use your Classify Study Fold to find an example of a food you eat that supplies four or more of the six nutrients. If there is none, which foods have the most nutrients?

Chapter ② Study Guide

Visualizing Main Ideas

Fill in the following table indicating good sources of vitamins and minerals.

Vitamin and Mineral Sources

Food Type	Source of Vitamin	Source of Mineral
Milk	D	
Spinach		iron
Meat		calcium, potassium
Eggs	E	
Carrots		sodium

Vocabulary Review

Vocabulary Words

a. amino acid
b. carbohydrate
c. chemical digestion
d. chyme
e. digestion
f. enzyme
g. fat
h. food group
i. mechanical digestion
j. mineral
k. nutrient
l. peristalsis
m. protein
n. villi
o. vitamin

THE PRINCETON REVIEW **Study Tip**

Practice reading graphs and charts. Make a table that contains the same information as a graph does. Have a friend make a graph of the information from a table. Exchange items and see if you can interpret each other's tables.

Using Vocabulary

In each sentence, replace the underlined word with the correct vocabulary word.

1. <u>Digestion</u> is the muscular contractions of the esophagus.

2. The <u>carbohydrates</u> increase the surface area of the small intestine.

3. The building blocks of proteins are <u>enzymes</u>.

4. The liquid product of digestion is called <u>villi</u>.

5. <u>Peristalsis</u> is the breakdown of food.

6. Your body's main source of energy is <u>vitamins</u>.

7. <u>Proteins</u> are inorganic nutrients.

8. Pears and apples belong to the same <u>mineral</u>.

9. <u>Chyme</u> is when food is chewed and mixed.

10. A <u>fat</u> is a nutrient needed in small quantities for growth and for regulating body functions.

Chapter 2 Assessment

Checking Concepts

Choose the word or phrase that best answers the question.

1. Where in humans does most chemical digestion occur?
   - **A)** duodenum
   - **B)** stomach
   - **C)** liver
   - **D)** large intestine

2. Which organ makes bile?
   - **A)** gallbladder
   - **B)** liver
   - **C)** stomach
   - **D)** small intestine

3. In which organ is water absorbed?
   - **A)** liver
   - **B)** esophagus
   - **C)** small intestine
   - **D)** large intestine

4. Which of these organs is an accessory organ?
   - **A)** mouth
   - **B)** stomach
   - **C)** small intestine
   - **D)** liver

5. What beneficial substances are produced by bacteria in the large intestine?
   - **A)** fats
   - **B)** minerals
   - **C)** vitamins
   - **D)** proteins

6. Which vitamin is found most abundantly in citrus fruits?
   - **A)** A
   - **B)** B
   - **C)** C
   - **D)** K

7. Where is hydrochloric acid added to the food mass?
   - **A)** mouth
   - **B)** stomach
   - **C)** small intestine
   - **D)** large intestine

8. From which food group should the largest number of servings in your diet come?
   - **A)** fruit
   - **B)** vegetable
   - **C)** milk, yogurt, and cheese
   - **D)** bread, cereal, rice, and pasta

9. Which food group contains yogurt and cheese?
   - **A)** dairy
   - **B)** grain
   - **C)** meat
   - **D)** fruit

10. Which organ produces enzymes that help in digestion of proteins, fats, and carbohydrates?
    - **A)** mouth
    - **B)** pancreas
    - **C)** large intestine
    - **D)** gallbladder

Thinking Critically

11. Food does not enter your body until it is absorbed into the blood. Explain why.

12. In what part of the digestive system do antacids work? Explain your choice.

13. Bile's action is similar to that of soap. Use this information to explain how bile works on fats.

14. Vitamin C and vitamin D are important for good health. Which of these might your body store? Explain your answer.

15. Based on your knowledge of food groups and nutrients, discuss the meaning of the familiar statement: "You are what you eat."

Developing Skills

16. **Making and Using Tables** In a table, sequence the order of organs in the digestive system through which food passes. Indicate whether ingestion, digestion, absorption, or elimination takes place.

17. **Comparing and Contrasting** Compare and contrast the three types of carbohydrates—sugar, starch, and fiber.

18. **Concept Mapping** Make a sequencing or events chain concept map showing the process of fat digestion.

19. **Classifying** Describe your favorite sandwich, then sort its parts into three of the nutrient categories—carbohydrates, proteins, and fats.

20. Making and Using Graphs Recommended Dietary Allowances (RDA) are the amounts of nutrients people need to maintain health. A product nutrient label is listed below. Make a bar graph of this information.

Recommended Dietary Allowances	
Nutrient	**Percent U.S. RDA**
Protein	2
Vitamin A	20
Vitamin C	25
Vitamin D	15
Calcium (Ca)	less than 2
Iron (Fe)	25
Zinc (Zn)	15
Total Fat	5
Saturated Fat	3
Cholesterol	0
Sodium	3

Performance Assessment

21. Project Research the ingredients used in antacid medications. Identify the compounds used to neutralize the excess stomach acid. Place an antacid tablet in a glass of vinegar. Using pH paper, check when the acid is neutralized. Record the time it took for the antacid to neutralize the vinegar. Repeat with different antacids.

 **Test Practice**

Kyle has just eaten a serving of vanilla ice cream after his lunch. The table below lists the nutritional facts of the ice cream.

Nutrition Facts of Vanilla Ice Cream		
Item	**Amount**	**DV (Daily Values)**
Serving Size	112 g	0
Calories	208	0
Total Fat	19 g	29%
Saturated Fat	11 g	55%
Cholesterol	0.125 g	42%
Sodium	0.90 g	4%
Total Carbohydrates	22 g	7%
Fiber	0 g	0%
Sugars	22 g	n/a
Protein	5 g	n/a
Calcium	0.117 g	15%
Iron	n/a	0%

Study the table and answer the following questions.

1. According to this information, which nutrient has the greatest Daily Value (DV) percentage?
A) total fat   **C)** saturated fat
B) calcium   **D)** cholesterol

2. According to the table, ice cream would NOT be a good source of _____ .
F) fiber   **H)** carbohydrates
G) saturated fat   **J)** calcium

Circulation

This interchange is simple compared to how blood travels within your body. In this chapter, you will discover how complex your circulatory system is. You'll learn what blood is made of and what it does for your body. You also will read about diseases that affect this body system. And you'll study a body system that helps protect you from disease.

What do you think?

Science Journal Look at the picture below. What do you think is happening? Here's a hint: *A part of your body is trying to protect itself.* Discuss your ideas with a classmate. Then, write your answer or best guess in your Science Journal.

EXPLORE ACTIVITY

I f you look at an aerial view of a road system, as shown in the photograph, you see roads leading in many directions. These roads provide a way to carry people and goods from one place to another. Your circulatory system is like a road system. Just as roads are used to transport goods to homes and factories, your blood vessels transport substances throughout your body.

Recognize transportation

1. Observe a map of your city, county, or state.
2. Identify roads that are interstates, as well as state and county routes, using the map key.
3. Plan a route to a destination that your teacher describes. Then plan a different return trip.
4. Draw a diagram in your Science Journal showing your routes to and from the destination.

Observe

If the destination represents your heart, what do the routes represent? In your Science Journal draw a comparison between a blocked road on your map and a clogged artery in your body.

Before You Read

FOLDABLES
Reading & Study Skills

Making a Concept Map Study Fold Make the following Foldable to help you organize information and diagram ideas about circulation.

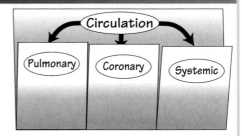

1. Place a sheet of paper in front of you so the long side is at the top. Fold the bottom of the paper to the top, stopping 4 cm from the top.
2. Fold both sides in to divide the paper into thirds and then unfold. Through the top thickness of paper, cut along each of the fold lines to form three tabs.
3. Draw an oval above the fold. Write *Circulation* inside the oval.
4. Draw three more ovals. Write the terms *Pulmonary*, *Coronary*, and *Systemic* as shown, and draw three arrows from the large oval to the smaller ovals.
5. As you read, write information about each system under its tab.

The Circulatory System

What **You'll Learn**

- **Compare and contrast** arteries, veins, and capillaries.
- **Explain** how blood moves through the heart.
- **Identify** the functions of the pulmonary and systemic circulation systems.

Vocabulary

atrium
ventricle
coronary circulation
pulmonary circulation
systemic circulation
artery
vein
capillary

Why **It's Important**

Your body's cells depend on the blood vessels to bring nutrients and remove wastes.

How Materials Move Through the Body

It's time to get ready for school, but your younger sister is taking a long time in the shower. "Don't use up all the water," you shout. Water is carried throughout your house in pipes that are part of the plumbing system. The plumbing system supplies water for all your needs and carries away wastes. Just as you expect water to flow when you turn on the faucet, your body needs a continuous supply of oxygen and nutrients and a way to remove wastes. In a similar way materials are moved throughout your body by your cardiovascular (kar dee oh VAS kyuh lur) system. It includes your heart, kilometers of blood vessels, and blood.

Blood vessels carry the blood to every part of your body, as shown in **Figure 1.** Blood moves oxygen and nutrients to cells and carries carbon dioxide and other wastes away from the cells. Sometimes the blood carries substances made in one part of the body to another part of the body where they are needed. Movement of materials into and out of your cells occurs by diffusion (dih FYEW zhun) and active transport. Diffusion occurs when a material moves from an area where there is more of it to an area where there is less of it. Active transport is the opposite of diffusion. Active transport requires energy, but diffusion does not.

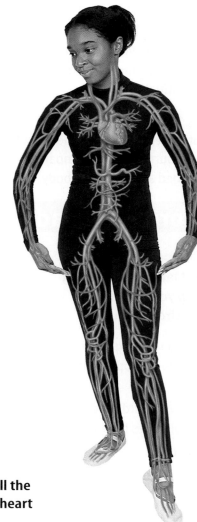

Figure 1
The blood is pumped by the heart to all the cells of the body and then back to the heart through a network of blood vessels.

The Heart

Your heart is an organ made of cardiac muscle tissue. It is located behind your breastbone, called the sternum, and between your lungs. Your heart has four compartments called chambers. The two upper chambers are called the right and left **atriums** (AY tree umz). The two lower chambers are called the right and left **ventricles** (VEN trih kulz). During one heartbeat, both atriums contract at the same time. Then, both ventricles contract at the same time. A one-way valve separates each atrium from the ventricle below it. The blood flows only in one direction from an atrium to a ventricle, then from a ventricle into a blood vessel. A wall prevents blood from flowing between the two atriums or the two ventricles. This wall keeps blood rich in oxygen separate from blood low in oxygen. If oxygen-rich blood and oxygen-poor blood were to mix, your body's cells would not get all the oxygen they need.

Scientists have divided the circulatory system into three sections—coronary circulation, pulmonary (PUL muh ner ee) circulation, and systemic circulation. The beating of your heart controls blood flow through each section.

Coronary Circulation Your heart has its own blood vessels that supply it with nutrients and oxygen and remove wastes. **Coronary** (KOR uh ner ee) **circulation,** as shown in **Figure 2,** is the flow of blood to and from the tissues of the heart. When the coronary circulation is blocked, oxygen and nutrients cannot reach all the cells of the heart. This can result in a heart attack.

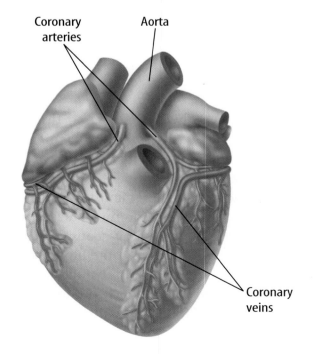

Coronary arteries

Aorta

Coronary veins

Figure 2
Like the rest of the body, the heart receives the oxygen and nutrients that it needs from the blood. The blood also carries away wastes from the heart's cells. On the diagram, you can see the coronary arteries, which nourish the heart.

A Blood, high in carbon dioxide and low in oxygen, returns from the body to the heart. It enters the right atrium through the superior and inferior vena cavae.

C Oxygen-rich blood travels from the lungs through the pulmonary vein and into the left atrium. The pulmonary veins are the only veins that carry oxygen-rich blood.

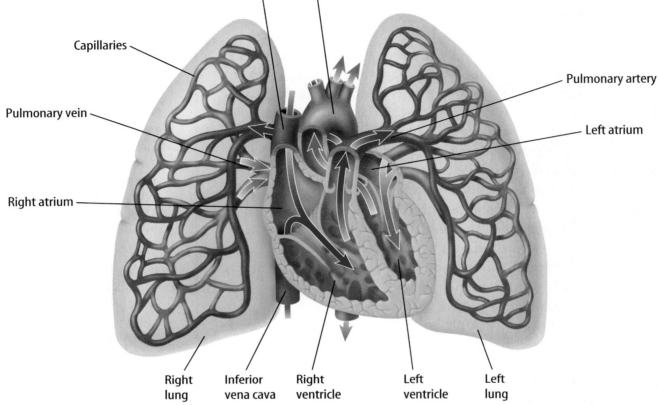

Capillaries

Pulmonary vein

Right atrium

Superior vena cava

Aorta

Pulmonary artery

Left atrium

Right lung

Inferior vena cava

Right ventricle

Left ventricle

Left lung

B The right atrium contracts, forcing the blood into the right ventricle. When the right ventricle contracts, the blood leaves the heart and goes through the pulmonary artery to the lungs. The pulmonary arteries are the only arteries that carry blood that is high in carbon dioxide.

D The left atrium contracts and forces the blood into the left ventricle. The left ventricle contracts, forcing the blood out of the heart and into the aorta.

Figure 3
Pulmonary circulation moves blood between the heart and lungs.

Pulmonary Circulation The flow of blood through the heart to the lungs and back to the heart is **pulmonary circulation**. Use **Figure 3** to trace the path blood takes through this part of the circulatory system. The blood returning from the body through the right side of the heart and to the lungs contains cellular wastes. The wastes include molecules of carbon dioxide and other substances. In the lungs, gaseous wastes diffuse out of the blood, and oxygen diffuses into the blood. Then the blood returns to the left side of the heart. In the final step of pulmonary circulation, the oxygen-rich blood is pumped from the left ventricle into the aorta (ay OR tuh), the largest artery in your body. Next, the oxygen-rich blood flows to all parts of your body.

Systemic Circulation Oxygen-rich blood moves to all of your organs and body tissues, except the heart and lungs, by **systemic circulation,** and oxygen-poor blood returns to the heart. Systemic circulation is the largest of the three sections of your circulatory system. **Figure 4** shows the major arteries (AR tuh reez) and veins (VAYNZ) of the systemic circulation system. Oxygen-rich blood flows from your heart in the arteries of this system. Then nutrients and oxygen are delivered by blood to your body cells and exchanged for carbon dioxide and wastes. Finally, the blood returns to your heart in the veins of the systemic circulation system.

✓ Reading Check *What are the functions of the systemic circulation system in your body?*

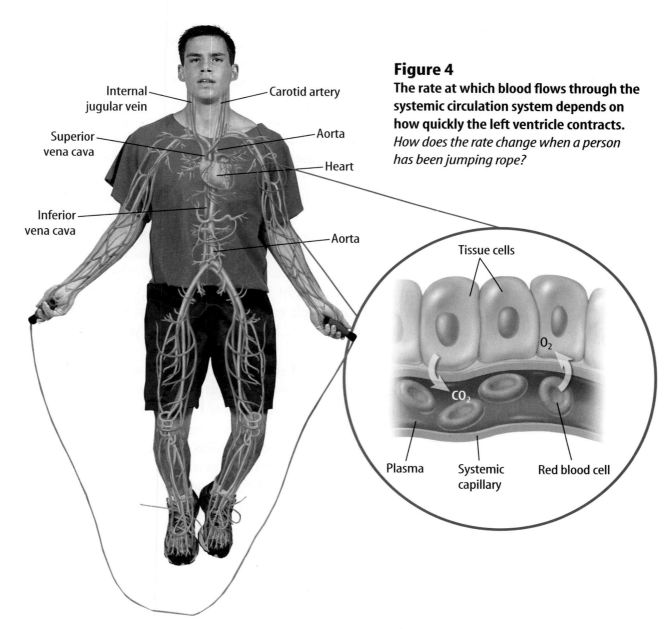

Figure 4
The rate at which blood flows through the systemic circulation system depends on how quickly the left ventricle contracts. *How does the rate change when a person has been jumping rope?*

Internal jugular vein

Carotid artery

Superior vena cava

Aorta

Heart

Inferior vena cava

Aorta

Tissue cells

O_2

CO_2

Plasma

Systemic capillary

Red blood cell

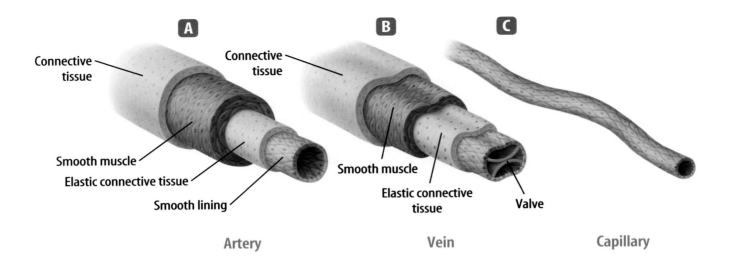

Connective tissue

Connective tissue

Smooth muscle

Elastic connective tissue

Smooth lining

Smooth muscle

Elastic connective tissue

Valve

Artery   **Vein**   **Capillary**

Figure 5
The structures of **A** arteries, **B** veins, and **C** capillaries are different. Valves in veins prevent blood from flowing backward. Capillaries are much smaller. Capillary walls are only one cell thick.

Blood Vessels

In the middle 1600s, scientists proved that blood moves in one direction in a blood vessel, like traffic on a one-way street. They discovered that blood moves by the pumping of the heart and flows from arteries to veins. But, they couldn't explain how blood gets from arteries to veins. Using a new invention of that time, the microscope, scientists discovered capillaries (KAP uh ler eez), the connection between arteries and veins.

Arteries As blood is pumped out of the heart, it travels through arteries, capillaries, and then veins. **Arteries** are blood vessels that carry blood away from the heart. Arteries, shown in **Figure 5A,** have thick, elastic walls made of connective tissue and smooth muscle tissue. Each ventricle of the heart is connected to an artery. The right ventricle is connected to the pulmonary artery, and the left ventricle is attached to the aorta. Every time your heart contracts, blood is moved from your heart into arteries.

Veins The blood vessels that carry blood back to the heart are called **veins,** as shown in **Figure 5B.** Veins have one-way valves that keep blood moving toward the heart. If blood flows backward, the pressure of the blood against the valves causes them to close. The flow of blood in veins also is helped by your skeletal muscles. When skeletal muscles contract, the veins in these muscles are squeezed and help blood move toward the heart. Two major veins return blood from your body to your heart. The superior vena cava returns blood from your head and neck. Blood from your abdomen and lower body returns through the inferior vena cava.

 **Reading Check** *What are the similarities and differences between arteries and veins?*

Capillaries Arteries and veins are connected by microscopic blood vessels called **capillaries,** as shown in **Figure 5C.** The walls of capillaries are only one cell thick. You can see capillaries when you have a bloodshot eye. They are the tiny red lines you see in the white area of your eye. Nutrients and oxygen diffuse into body cells through the thin capillary walls. Waste materials and carbon dioxide diffuse from body cells into the capillaries.

Blood Pressure

If you fill a balloon with water and then push on it, the pressure moves through the water in all directions, as shown in **Figure 6.** Your circulatory system is like the water balloon. When your heart pumps blood through the circulatory system, the pressure of the push moves through the blood. The force of the blood on the walls of the blood vessels is called blood pressure. This pressure is highest in arteries and lowest in veins. When you take your pulse, you can feel the waves of pressure. This rise and fall of pressure occurs with each heartbeat. Normal resting pulse rates are 60 to 100 heartbeats per minute for adults, and 80 to 100 beats per minute for children.

Measuring Blood Pressure Blood pressure is measured in large arteries and is expressed by two numbers, such as 120 over 80. The first number is a measure of the pressure caused when the ventricles contract and blood is pushed out of the heart. This is called the systolic (sihs TAHL ihk) pressure. Then, blood pressure drops as the ventricles relax. The second number is a measure of the diastolic (di uh STAHL ihk) pressure that occurs as the ventricles fill with blood just before they contract again.

Controlling Blood Pressure Your body tries to keep blood pressure normal. Special nerve cells in the walls of some arteries sense changes in blood pressure. When pressure is higher or lower than normal, messages are sent to your brain by these nerve cells. Then messages are sent by your brain to raise or lower blood pressure—by speeding up or slowing the heart rate for example. This helps keep blood pressure constant within your arteries. When blood pressure is constant, enough blood reaches all organs and tissues in your body and delivers needed nutrients to every cell.

Some molecules of nutrients are forced through capillary walls by the force of blood pressure. What is the cause of the pressure? Discuss your answer with a classmate. Then write your answer in your Science Journal.

Figure 6
When pressure is exerted on a fluid in a closed container, the pressure is transmitted through the liquid in all directions. Your circulatory system is like a closed container.

Water-filled balloon

Figure 7

Healthy blood vessels have smooth, unobstructed interiors like the one at the right. Atherosclerosis is a disease in which fatty substances build up in the walls of arteries, such as the coronary arteries that supply the heart muscle with oxygen-rich blood. As illustrated below, these fatty deposits can gradually restrict—and ultimately block—the life-giving river of blood that flows through an artery.

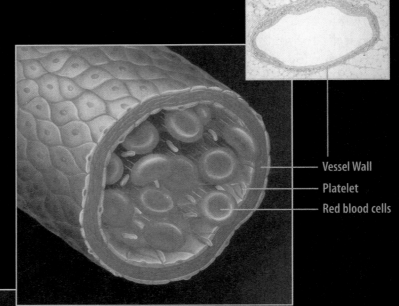

Vessel Wall
Platelet
Red blood cells

Vessel Wall

Plaque

▲ **HEALTHY ARTERY** The illustration and photo above show a normal functioning artery.

◄ **PARTIALLY CLOGGED ARTERY** The illustration and inset photo at left show fatty deposits, called plaques, that have formed along the artery's inner wall. As the diagram illustrates, plaques narrow the pathway through the artery, restricting and slowing blood flow. As blood supply to the heart muscle cells dwindles, they become starved for oxygen and nutrients.

▶ **NEARLY BLOCKED ARTERY** In the illustration and photo at right, fatty deposits have continued to build. The pathway through the coronary artery has gradually narrowed until blood flow is very slow and nearly blocked. Under these conditions, the heart muscle cells supplied by the artery are greatly weakened. If blood flow stops entirely, a heart attack will result.

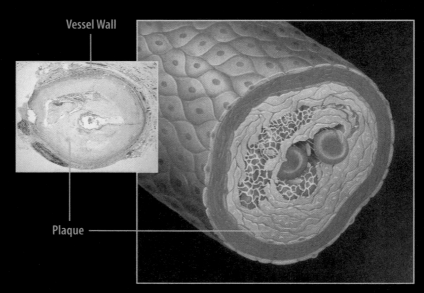

Vessel Wall

Plaque

Cardiovascular Disease

Any disease that affects the cardiovascular system—the heart, blood vessels, and blood—can seriously affect the health of your entire body. People often think of cancer and automobile accidents as the leading causes of death in the United States. However, heart disease is the leading cause of death.

Atherosclerosis One leading cause of heart disease is called atherosclerosis (ah thur oh skluh ROH sus). In this condition, shown in **Figure 7,** fatty deposits build up on arterial walls. Eating foods high in cholesterol and saturated fats can cause these deposits to form. Atherosclerosis can occur in any artery in the body, but deposits in coronary arteries are especially serious. If a coronary artery is blocked, a heart attack can occur. Open heart surgery may then be needed to correct the problem.

Hypertension Another condition of the cardiovascular system is called hypertension (HI pur ten chun), or high blood pressure. **Figure 8** shows the instruments used to measure blood pressure. When blood pressure is higher than normal most of the time, extra strain is placed on the heart. The heart must work harder to keep blood flowing. One cause of hypertension is atherosclerosis. A clogged artery can increase pressure within the vessel. The walls become stiff and hard, like a metal pipe. The artery walls no longer contract and dilate easily because they have lost their elasticity.

Heart Failure Heart failure results when the heart cannot pump blood efficiently. It might be caused when heart muscle tissue is weakened by disease or when heart valves do not work properly. When the heart does not pump blood properly, fluids collect in the arms, legs, and lungs. People with heart failure usually are short of breath and tired.

✔ **Reading Check** *What is heart failure?*

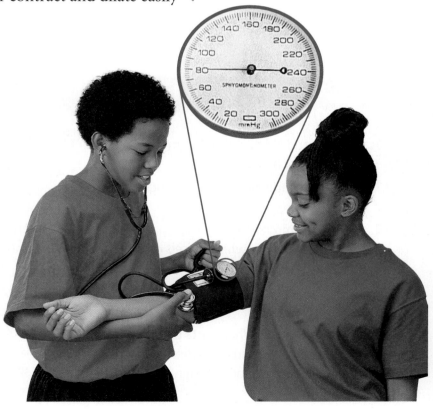

Figure 8
Blood pressure is measured in large arteries using a blood pressure cuff and stethoscope.

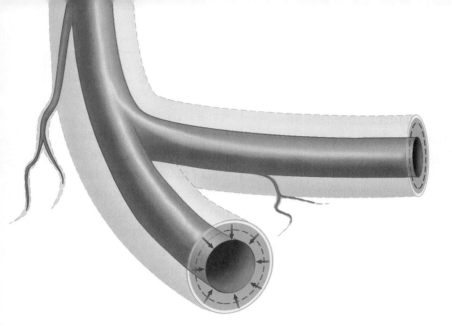

Figure 9
Nicotine, present in tobacco, contracts blood vessels and causes the body to release hormones that raise blood pressure.

Preventing Cardiovascular Disease Having a healthy lifestyle is important for the health of your cardiovascular system. The choices you make to maintain good health may reduce your risk of future serious illness. Regular checkups, a healthful diet, and exercise are part of a heart-healthy lifestyle.

Many diseases, including cardiovascular disease, can be prevented by following a good diet. Choose foods that are low in salt, sugar, cholesterol, and saturated fats. Being overweight is associated with heart disease and high blood pressure. Large amounts of body fat force the heart to pump faster.

Learning to relax and having a regular program of exercise can help prevent tension and relieve stress. Exercise also strengthens the heart and lungs, helps in controlling cholesterol, tones muscles, and helps lower blood pressure.

Another way to prevent cardiovascular disease is to not smoke. Smoking causes blood vessels to contract, as shown in **Figure 9,** and makes the heart beat faster and harder. Smoking also increases carbon monoxide levels in the blood. Not smoking helps prevent heart disease and a number of respiratory system problems, too.

Section 1 Assessment

1. Compare and contrast the structure of the three types of blood vessels.
2. Explain the pathway of blood through the heart.
3. Contrast pulmonary and systemic circulation. Identify which vessels carry oxgen-rich blood.
4. Explain how exercise can help prevent heart disease.
5. **Think Critically** What waste product builds up in blood and cells when the heart is unable to pump blood efficiently?

Skill Builder Activities

6. **Concept Mapping** Make an events chain concept map to show pulmonary circulation beginning at the right atrium and ending at the aorta. **For more help, refer to the** Science Skill Handbook.
7. **Using a Database** Research diseases of the circulatory system. Make a database showing what part of the circulatory system is affected by each disease. Categories should include the organs and vessels of the circulatory system. **For more help, refer to the** Technology Skill Handbook.

Activity

The Heart as a Pump

The heart is a pumping organ. Blood is forced through the arteries as heart muscles contract and then relax. This creates a series of waves in blood as it flows through the arteries. These waves are called the pulse. Try this activity to learn how physical activity affects your pulse.

What You'll Investigate
What does the pulse rate tell you about the work of the heart?

Materials
watch or clock with a second hand
*stopwatch
*Alternate materials

Goals
- ■ **Observe** pulse rate.
- ■ **Compare** pulse rate at rest to rate after jogging.

Data Table

Pulse Rate		
Pulse Rate	Partner's	Yours
At Rest		
After Jogging		

Procedure
1. Make a table like the one shown. Use it to record your data.
2. Sit down to take your pulse. Your partner will serve as the recorder.
3. Find your pulse by placing your middle and index fingers over the radial artery in your wrist as shown in the photo.
   WARNING: *Do not press too hard.*
4. **Count** each beat of the radial pulse silently for 15 s. Multiply the number of beats by four to find your pulse rate per minute. Have your partner record the number in the data table.
5. Now jog in place for 1 min and take your pulse again. Count the beats for 15 s.
6. **Calculate** this new pulse rate and have your partner record it in the data table.
7. Reverse roles with your partner and repeat steps 2 through 6.
8. **Collect** and record the new data.

Conclude and Apply
1. How does the pulse rate change?
2. What causes the pulse rate to change?
3. What can you infer about the heart as a pumping organ?

Communicating Your Data

Record the class average for pulse rate at rest and after jogging. Compare the class averages to your data. **For more help, refer to the** Science Skill Handbook.

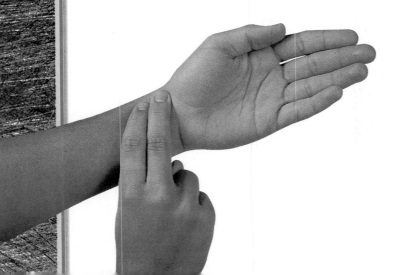

Blood

As You Read

What You'll Learn

■ **Identify** the parts and functions of blood.
■ **Explain** why blood types are checked before a transfusion.
■ **Give examples** of diseases of blood.

Vocabulary

plasma
hemoglobin
platelet

Why It's Important

Blood plays a part in every major activity of your body.

Functions of Blood

You take a last, deep, calming breath before plunging into a dark, vessel-like tube. Water is everywhere. You take a hard right turn, then left as you streak through a narrow tunnel of twists and turns. The water transports you down the slide much like the way blood carries substances to all parts of your body. Blood has four important functions.

1. Blood carries oxygen from your lungs to all your body cells. Carbon dioxide diffuses from your body cells into your blood. Your blood carries carbon dioxide to your lungs to be exhaled.

2. Blood carries waste products from your cells to your kidneys to be removed.

3. Blood transports nutrients and other substances to your body cells.

4. Cells and molecules in blood fight infections and help heal wounds.

Anything that disrupts or changes these functions affects all the tissues of your body. Can you understand why blood is sometimes called the tissue of life?

Parts of Blood

A close look at blood tells you that blood is not just a red-colored liquid. Blood is a tissue made of plasma (PLAZ muh), red and white blood cells, and platelets (PLAYT luts), as shown in **Figure 10.** Blood makes up about eight percent of your body's total mass. If you weigh 45 kg, you have about 3.6 kg of blood moving through your body. The amount of blood in an adult would fill five 1-L bottles. If this volume decreases rapidly because of an injury or disease, blood pressure will fall and the body may go into shock.

Plasma The liquid part of blood, which is made mostly of water is called **plasma.** It makes up more than half the volume of blood. Nutrients, minerals, and oxygen are dissolved in plasma so that they can be carried to body cells. Wastes from body cells are also carried in plasma.

Figure 10
The blood in this graduated cylinder has separated into its parts. Each part plays a key role in body functions.

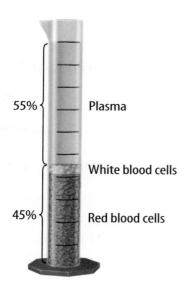

55% — Plasma

White blood cells

45% — Red blood cells

Blood Cells A cubic millimeter of blood has about 5 million red blood cells. These disk-shaped blood cells, shown in **Figure 11,** are different from other cells in your body because they have no nuclei. They contain **hemoglobin** (HEE muh gloh bun), which is a molecule that carries oxygen and carbon dioxide. Hemoglobin carries oxygen from your lungs to your body cells. Then it carries some of the carbon dioxide from your body cells back to your lungs. The rest of the carbon dioxide is carried in the cytoplasm of red blood cells and in plasma. Red blood cells have a life span of about 120 days. They are made at a rate of 2 million to 3 million per second in the center of long bones like the femur in your thigh. Red blood cells wear out and are destroyed at about the same rate.

In contrast to red blood cells, a cubic millimeter of blood has about 5,000 to 10,000 white blood cells. White blood cells fight bacteria, viruses, and other invaders of your body. Your body reacts to invaders by increasing the number of white blood cells. These cells leave the blood through capillary walls and go into the tissues that have been invaded. Here, they destroy bacteria and viruses and absorb dead cells. The life span of white blood cells varies from a few days to many months.

Circulating with the red and white blood cells are platelets. **Platelets** are irregularly shaped cell fragments that help clot blood. A cubic millimeter of blood can contain as many as 400,000 platelets. Platelets have a life span of five to nine days.

SCIENCE
Online

Research Visit the Glencoe Science Web site at **science.glencoe.com** for more information about the types of human white blood cells and their functions. Communicate to your class what you learn.

Figure 11
Red blood cells supply your body with oxygen, and white blood cells and platelets have protective roles.

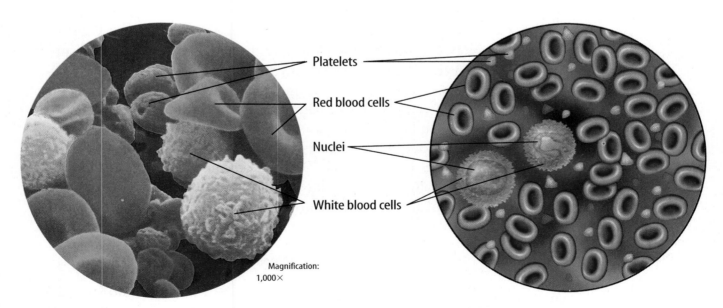

Platelets

Red blood cells

Nuclei

White blood cells

Magnification: 1,000×

A Platelets help stop bleeding. Platelets not only plug holes in small vessels, they also release chemicals that help form filaments of fibrin.

B Several types, sizes, and shapes of white blood cells exist. These cells destroy bacteria, viruses, and foreign substances.

Figure 12

A When the skin is damaged, a sticky blood clot seals the leaking blood vessel. **B** Eventually, a scab forms to protect the wound from further damage and allow it to heal.

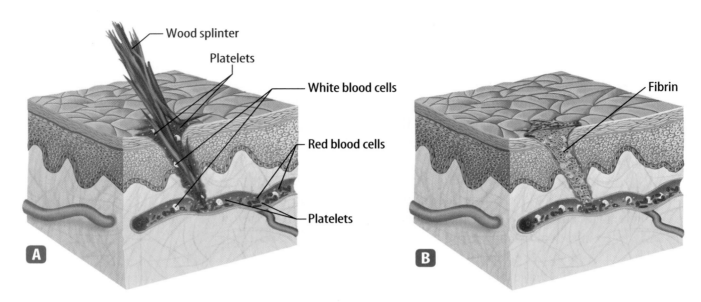

- Wood splinter
- Platelets
- White blood cells
- Red blood cells
- Platelets
- Fibrin

A

B

Mini LAB

Modeling Scab Formation

Procedure 🥽

1. Place a 5-cm × 5-cm square of **gauze** on a piece of **aluminum foil.**
2. Place several drops of a **liquid bandage solution** onto the gauze and let it dry. Keep the liquid bandage away from eyes and mouth.
3. Use a **dropper** to place one drop of **water** onto the area of the liquid bandage. Place another drop of water in another area of the gauze.

Analysis

1. Compare the drops of water in both areas.
2. Describe how the treated area of the gauze is like a scab.

Blood Clotting

You're running with your dog in a park, when all of a sudden you trip and fall down. Your knee starts to bleed, but the bleeding stops quickly. Already the wounded area has begun to heal. Bleeding stops because platelets and clotting factors in your blood make a blood clot that plugs the wounded blood vessels. A blood clot also acts somewhat like a bandage. When you cut yourself, platelets stick to the wound and release chemicals. Then substances called clotting factors carry out a series of chemical reactions. These reactions cause threadlike fibers called fibrin (FI brun) to form a sticky net, as shown in **Figure 12.** This net traps escaping blood cells and plasma and forms a clot. The clot helps stop more blood from escaping. After the clot is in place and becomes hard, skin cells begin the repair process under the scab. Eventually, the scab is lifted off. Bacteria that might get into the wound during the healing process are destroyed by white blood cells.

✔ **Reading Check** *What blood components help form blood clots?*

Most people will not bleed to death from a minor wound, such as a cut or scrape. However, some people have a genetic condition called hemophilia (hee muh FIHL ee uh). Their plasma lacks one of the clotting factors that begins the clotting process. A minor injury can be a life threatening problem for a person with hemophilia.

Blood Types

Blood clots stop blood loss quickly in a minor wound, but with a serious wound a person might lose a lot of blood. A blood transfusion might be necessary. During a blood transfusion, a person receives donated blood or parts of blood. The medical provider must be sure that the right type of blood is given. If the wrong type is given, the red blood cells will clump together. Then, clots form in the blood vessels and the person could die.

Table 1 Blood Types		
Blood Type	**Antigen**	**Antibody**
A	A	Anti-B
B	B	Anti-A
AB	A, B	None
O	None	Anti-A Anti-B

The ABO Identification System People can inherit one of four types of blood: A, B, AB, or O, as shown in **Table 1.** Types A, B, and AB have chemical identification tags called antigens (AN tih junz) on their red blood cells. Type O red blood cells have no antigens.

Each blood type also has specific antibodies in its plasma. Antibodies are proteins that destroy or neutralize substances that do not belong in or are not part of your body. Because of these antibodies, certain blood types cannot be mixed. This limits blood transfusion possibilities as shown in **Table 2.** If type A blood is mixed with type B blood, the antibodies in type A blood determine that type B blood does not belong there. The antibodies in type A blood cause the type B red blood cells to clump. In the same way, type B blood antibodies cause type A blood to clump. Type AB blood has no antibodies, so people with this blood type can receive blood from A, B, AB, and O types. Type O blood has both A and B antibodies.

Table 2 Blood Transfusion Possibilities		
Type	**Can Receive**	**Can Donate To**
A	O, A	A, AB
B	O, B	B, AB
AB	all	AB
O	O	all

Do you know what to do in an emergency? To learn about first aid, see the **Emergencies Field Guide** at the back of the book.

✔ **Reading Check**

Why are people with type O blood called universal donors?

The Rh Factor Another chemical identification tag in blood is the Rh factor. The Rh factor also is inherited. If the Rh factor is on red blood cells, the person has Rh-positive (Rh+) blood. If it is not present, the person's blood is called Rh-negative (Rh−). If an Rh− person receives a blood transfusion from an Rh+ person, he or she will produce antibodies against the Rh factor. These antibodies can cause Rh+ cells to clump. Clots then form in the blood vessels and the person could die.

When an Rh− mother is pregnant with an Rh+ baby, the mother might make antibodies to the child's Rh factor. Close to the time of birth, Rh antibodies from the mother can pass from her blood vessels into the baby's blood vessels. These antibodies can destroy the baby's red blood cells. If this happens, the baby must receive a blood transfusion before or right after birth. At 28 weeks of pregnancy and immediately after the birth, an Rh− mother can receive an injection that blocks the production of antibodies to the Rh+ factor. These injections prevent this life-threatening situation from occurring in future pregnancies. To prevent deadly results, blood groups and Rh factor are checked before transfusions and during pregnancies.

Reading Check *Why is it important to check Rh factor?*

Artificial blood substances have been developed to use in blood transfusions. They can carry oxygen and carbon dioxide. Predict what other properties they must have to be safe. Write your prediction in your Science Journal.

Problem-Solving Activity

Will there be enough blood donors?

Successful human blood transfusions began during World War II. This practice is much safer today due to extensive testing of the donated blood prior to transfusion. Health care professionals have determined that each blood type can receive certain other blood types as illustrated in **Table 2.**

Blood Type Distribution		
	Rh + (%)	**Rh − (%)**
O	37	7
A	36	6
B	9	1
AB	3	1

Identifying the Problem

The table on the right lists the average distribution of blood types in the United States. The data are recorded as percents, or a sample of 100 people. By examining these data and the data in **Table 2,** can you determine safe donors for each blood type? Recall that people with Rh− blood cannot receive a transfusion from an Rh+ donor.

Solving the Problem

1. If a Type B, Rh+ person needs a blood transfusion, how many possible donors are there?
2. Frequently, the supply of donated blood runs low. Which blood type and Rh factor would be most affected in such a shortage? Explain your answer.

Diseases of Blood

Because blood circulates to all parts of your body and performs so many important functions, any disease of the blood is a cause for concern. One common disease of the blood is anemia (uh NEE mee uh). In this disease of red blood cells, body tissues can't get enough oxygen and are unable to carry on their usual activities. Anemia has many causes. Sometimes, anemia is caused by the loss of large amounts of blood. A diet lacking iron or certain vitamins also might cause anemia. In addition, anemia can be the result of another disease or a side effect of treatment for a disease. Still other types of anemia are inherited problems related to the structure of the red blood cells. Cells from one such type of anemia, sickle-cell anemia, are shown in **Figure 13.**

Leukemia (lew KEE mee uh) is a disease in which one or more types of white blood cells are made in excessive numbers. These cells are immature and do not fight infections well. These immature cells fill the bone marrow and crowd out the normal cells. Then not enough red blood cells, normal white blood cells, and platelets can be made. Some types of leukemia affect children. Other kinds are more common in adults. Medicines, blood transfusions, and bone marrow transplants are used to treat this disease. If the treatments are not successful, the person will eventually die from related complications.

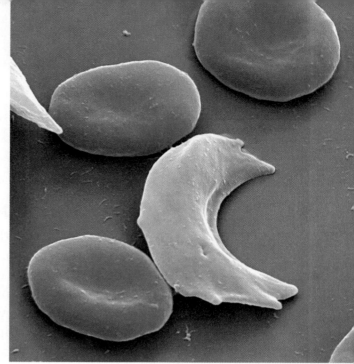

Magnification: 1,500×

Figure 13
Persons with sickle-cell anemia have misshapened red blood cells. The sickle-shaped cells clog the capillaries of a person with this disease. Oxygen cannot reach tissues served by the capillaries, and wastes cannot be removed. *How does this damage the affected tissues?*

Section Assessment

1. What are the four functions of blood in the body?

2. Compare red blood cells, white blood cells, and platelets.

3. Why are blood type and Rh factor checked before a transfusion?

4. Describe two diseases of blood.

5. **Think Critically** Think about the main job of your red blood cells. If red blood cells couldn't deliver oxygen to your cells, what would be the condition of your body tissues?

Skill Builder Activities

6. **Interpreting Data** Look at the data in **Table 2** about blood group interactions. To which group(s) can blood type AB donate blood? **For more help, refer to the** Science Skill Handbook.

7. **Using Percentages** Find the total number of red blood cells, white blood cells, and platelets in 1 mm^3 of blood. Calculate what percentage of the total each type is. **For more help, refer to the** Math Skill Handbook.

③ The Lymphatic System

As You Read

What You'll Learn

- **Describe** functions of the lymphatic system.
- **Identify** where lymph comes from.
- **Explain** how lymph organs help fight infections.

Vocabulary

lymph
lymphatic system
lymphocyte
lymph node

Why It's Important

The lymphatic system helps protect you from infections and diseases.

Functions of the Lymphatic System

You're thirsty after a long walk home in the hot sun. You turn on the water faucet and fill a glass with water. The excess water runs down the drain. In a similiar way, your body's tissue fluid is removed by the lymphatic (lihm FAT ihk) system. The nutrient, water, and oxygen molecules in blood diffuse through capillary walls to nearby cells. Water and other substances become part of the tissue fluid that is found between cells. This fluid is collected and returned to the blood by the lymphatic system.

After tissue fluid diffuses into the lymphatic capillaries it is called **lymph** (LIHMF). Your **lymphatic system,** as shown in **Figure 14,** carries lymph through a network of lymph capillaries and larger lymph vessels. Then, the lymph drains into large veins near the heart. No heartlike structure pumps the lymph through the lymphatic system. The movement of lymph depends on the contraction of smooth muscles in lymph vessels and skeletal muscles. Lymphatic vessels, like veins, have valves that keep lymph from flowing backward.

In addition to water and dissolved substances, lymph also contains **lymphocytes** (LIHM fuh sites), a type of white blood cell. Lymphocytes help your body defend itself against disease-causing organisms. If the lymphatic system is not working properly, severe swelling occurs because the tissue fluid cannot get back to the blood.

✓ Reading Check *What are the differences and similarities between lymph and blood?*

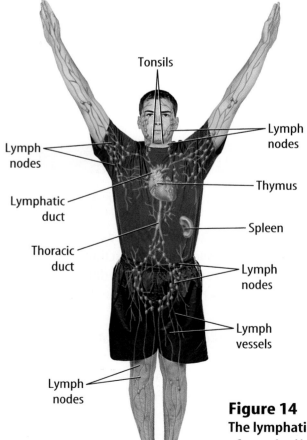

Tonsils
Lymph nodes
Lymph nodes
Lymphatic duct
Thymus
Spleen
Thoracic duct
Lymph nodes
Lymph vessels
Lymph nodes

Figure 14
The lymphatic system is connected by a network of vessels. *How do muscles help move lymph?*

Lymphatic Organs

Before lymph enters the blood, it passes through lymph nodes, which are bean-shaped organs of varying sizes found throughout the body. **Lymph nodes** filter out microorganisms and foreign materials that have been taken up by lymphocytes. When your body fights an infection, lymphocytes fill the lymph nodes. The lymph nodes become warm, reddened, and tender to the touch. After the invaders are destroyed, the redness, warmth, and tenderness in the lymph nodes go away.

Besides lymph nodes, three important lymphatic organs are the tonsils, the thymus, and the spleen. Tonsils are in the back of your throat. They protect your body from harmful microorganisms that enter through your mouth and nose. Your thymus is a soft mass of tissue located behind the sternum. It makes lymphocytes that travel to other lymph organs. The spleen is the largest lymphatic organ. It is located behind the upper-left part of the stomach and filters the blood by removing worn out and damaged red blood cells. Cells in the spleen take up and destroy bacteria and other substances that invade your body.

A Disease of the Lymphatic System

As you probably have heard, HIV is a deadly virus. When HIV enters a person's body, it attacks and destroys a certain kind of lymphocyte called helper T-cells that help make antibodies to fight infections. This affects a person's immunity to certain diseases. Usually, the person dies from these other infections, not from the HIV infection.

SCIENCE *Online*

Data Update For an online update about HIV and AIDS, visit the Glencoe Science Web site at **science.glencoe.com** and select the appropriate chapter.

Section 3 Assessment

1. List the organs of your lymphatic system and describe their functions.
2. Where does lymph come from and how does it get into the lymphatic capillaries?
3. How do lymphatic organs fight infection?
4. What events occur when HIV enters the body?
5. **Think Critically** When the amount of fluid in the spaces between cells increases, so does the pressure in these spaces. What do you infer will happen?

Skill Builder Activities

6. **Concept Mapping** The circulatory system and the lymphatic system work together in several ways. Make a concept map comparing the two systems. **For more help, refer to the** Science Skill Handbook.
7. **Communicating** An infectious microorganism enters your body. In your Science Journal, describe how the lymphatic system protects the body against the microorganism. **For more help, refer to the** Science Skill Handbook.

Blood Type Reactions

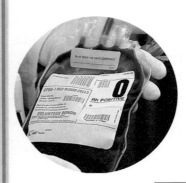

Human blood can be classified into four main blood types——A, B, AB, and O. These types are determined by the presence or absence of antigens on the red blood cells. After blood is collected into a transfusion bag, it is tested to determine the blood type. The type is labeled clearly on the bag. Blood is refrigerated to keep it fresh and available for transfusion.

Recognizing the Problem

What happens when two different blood types are mixed?

Forming a Hypothesis

Based on your reading and observations, state a hypothesis about how different blood types will react to each other.

Safety Precautions 🥽 🧤 🚫 🔥

Do not taste, eat, or drink any materials used in the lab.

Possible Materials

simulated blood (10 mL low-fat milk and 10 mL water plus red food coloring)

lemon juice as antigen A (for blood types B and O)

water as antigen A (for blood types A and AB)

droppers

small paper cups

marking pen

10-mL graduated cylinder

Goals

■ **Design** an experiment that simulates the reactions between different blood types.

■ **Identify** which blood types can donate to which other blood types.

Test Your Hypothesis

Plan

1. As a group, agree upon the hypothesis and decide how you will test it. Identify the results that will confirm the hypothesis.
2. **List** the steps you must take and the materials you will need to test your hypothesis. Be specific. Describe exactly what you will do in each step.
3. **Prepare** a data table like the one at the right in your Science Journal to record your observations.
4. Reread the entire experiment to make sure all steps are in logical order.
5. **Identify** constants and variables. Blood type O will be the control.

Blood Type Reactions	
Blood Type	Clumping (Yes or No)
A	
B	
AB	
O	

Do

1. Make sure your teacher approves your plan before you start.
2. Carry out the experiment according to the approved plan.
3. While doing the experiment, record your observations and complete the data table in your Science Journal.

Analyze Your Data

1. **Compare** the reactions of each blood type (A, B, AB, and O) when antigen A was added to the blood.
2. **Observe** where clumping took place.
3. **Compare** your results with those of other groups.
4. What was the control factor in this experiment?
5. What were your variables?

Draw Conclusions

1. Did the results support your hypothesis? Explain.
2. **Predict** what might happen to a person if other antigens are not matched properly.
3. What would happen in an investigation with antigen B added to each blood type?

Communicating Your Data

Write a brief report on how blood is tested to determine blood type. **Describe** why this is important to know before receiving a blood transfusion. **For more help, refer to the** Science Skill Handbook.

TIME

SCIENCE AND HISTORY

**SCIENCE
CAN CHANGE
THE COURSE
OF HISTORY**

Dr. Daniel Hale Williams was a pioneer in open-heart surgery.

Have a Heart

People didn't always know where blood came from or how it moved through the body

"Ouch!" You prick your finger, and when blood starts to flow out of the cut, you put on a bandage. But if you were a scientist living long ago, you might have also asked yourself some questions: How did your blood get to the tip of your finger? And why and how does it flow through (and sometimes out of!) your body?

As early as the 1500s, a Spanish scientist named Miguel Serveto (mee GEL ● ser VET oh) asked that question. His studies led him to the theory that blood circulated throughout the human body, but he didn't know how or why.

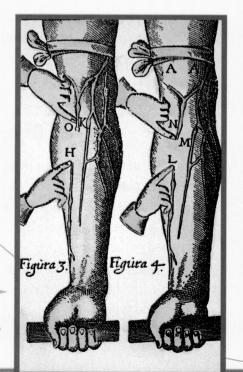

A woodcut from William Harvey's book demonstrates his theory of blood circulation. The book was published in 1628.

William Harvey (standing) explains his theory of blood circulation to England's King Charles I. The boy is the future King Charles II.

About 100 years later, William Harvey, an English doctor, explored Servet's idea. Harvey studied animals to develop a theory about how the heart and the circulatory system work. Back then, most people thought that food was turned into blood by the liver.

Harvey knew this was untrue from his observations of animals. Blood was pumped from the heart throughout the body, Harvey hypothesized. Then it returned to the heart and recirculated. He published his ideas in 1628 in his famous book, *On the Motion of the Heart and Blood in Animals*. His theories were correct, but many of Harvey's patients left him. They thought his ideas were bloody ridiculous! But over time, Harvey's book became the basis for all modern research on heart and blood vessels.

Medical Pioneer

More than two centuries later, another pioneer would step forward and use Harvey's ideas to change the science frontier again. His name is Dr. Daniel Hale Williams. In 1893, Williams used what he knew about the heart and blood circulation to become a new kind of medical pioneer. He performed the first open-heart surgery by removing a knife from the heart of a stabbing victim. He stitched the wound to the fluid sac surrounding the heart, and the patient lived for several years afterward. In 1970, the U.S. recognized this American medical pioneer by issuing a stamp in his honor.

CONNECTIONS Report Pioneers in medicine continue to help people lead longer lives. Identify a pioneer in science or medicine who has changed our lives for the better. Find out how this person started in the field, and how they came to make an important discovery. Give a presentation to the class.

SCIENCE *Online*

For more information, visit science.glencoe.com

Chapter **3** Study Guide

Reviewing Main Ideas

Section 1 The Circulatory System

1. Arteries carry blood away from the heart. Capillaries allow the exchange of nutrients, oxygen, and wastes in cells. Veins return blood to the heart.

2. Blood that is high in carbon dioxide enters the right atrium, moves to the right ventricle, and then goes to the lungs through the pulmonary artery. Oxygen-rich blood returns to the left atrium, moves to the left ventricle, and then leaves through the aorta.

3. Pulmonary circulation is the path of blood between the heart and lungs. Circulation through the rest of the body is called systemic circulation. Coronary circulation is the flow of blood to tissues of the heart. *Trace pulmonary circulation in this image.*

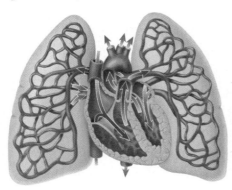

4. A healthy lifestyle is important for the health of your cardiovascular system.

Section 2 Blood

1. Red blood cells carry oxygen and carbon dioxide, platelets form clots, and white blood cells fight infection. Plasma carries nutrients, blood cells, and other substances.

2. A, B, AB, and O blood types are determined by the presence or absence of antigens on red blood cells.

3. Anemia is a disease of red blood cells, shown in the photograph, in which not enough oxygen is carried to the body's cells. *What are some causes of anemia?*

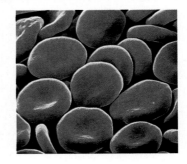

4. Leukemia is a disease of white blood cells in which one or more types of white blood cells are present in excessive numbers. These cells are immature and do not fight infection well.

Section 3 The Lymphatic System

1. Lymph structures filter blood, produce white blood cells that destroy bacteria and viruses, and destroy worn out blood cells. *How does this lymph node help your body fight disease?*

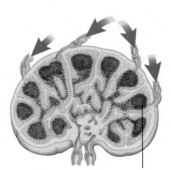

Lymphocytes

2. HIV attacks helper T-cells, which are a type of lymphocyte. The person is unable to fight infections well.

FOLDABLES
Reading & Study Skills

After You Read

Add a labeled arrow and an oval under each of the following and write the major organ of that part of the circulatory system: *Pulmonary*, *Coronary*, and *Systemic*.

Visualizing Main Ideas

Fill in the concept map on the functions of the parts of the blood.

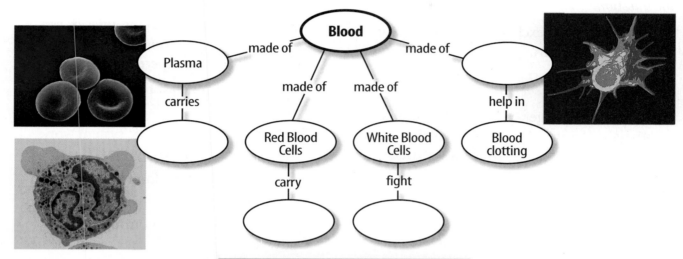

Vocabulary Review

Vocabulary Words

a. artery
b. atrium
c. capillary
d. coronary circulation
e. hemoglobin
f. lymph
g. lymph node
h. lymphatic system
i. lymphocyte
j. plasma
k. platelet
l. pulmonary circulation
m. systemic circulation
n. vein
o. ventricle

Study Tip

Make a study schedule for yourself. If you have a planner, write down exactly which hours you plan to spend studying and stick to it.

Using Vocabulary

Replace the underlined words with the correct vocabulary words.

1. The plasma carries blood to the heart.

2. The ventricle transports tissue fluid through a network of vessels.

3. Lymph is the chemical in red blood cells.

4. Lymph nodes are cell fragments.

5. The smallest blood vessels are called the lymphatic system.

6. The flow of blood to and from the lungs is called coronary circulation.

7. Hemoglobin helps protect your body against infections.

8. The largest section of the circulatory system is the pulmonary circulation.

9. Lymphocytes are blood vessels that carry blood away from the heart.

10. The two lower chambers of the heart are called the right and left atriums.

Chapter **3** Assessment

Checking Concepts

Choose the word or phrase that best answers the question.

1. Where does the exchange of food, oxygen, and wastes occur?
   A) arteries   C) veins
   B) capillaries   D) lymph vessels

2. Where does oxygen-rich blood enter first?
   A) right atrium   C) left ventricle
   B) left atrium   D) right ventricle

3. What is circulation to all body organs called?
   A) coronary   C) systemic
   B) pulmonary   D) organic

4. Where is blood under greatest pressure?
   A) arteries   C) veins
   B) capillaries   D) lymph vessels

5. Which of these is a function of blood?
   A) digest food   C) dissolve bone
   B) produce CO_2 D) carry oxygen

6. Which cells fight off infection?
   A) red blood   C) white blood
   B) bone   D) nerve

7. Of the following, which carries oxygen in blood?
   A) red blood cells   C) white blood cells
   B) platelets   D) lymph

8. What is required to clot blood?
   A) plasma   C) platelets
   B) oxygen   D) carbon dioxide

9. What kind of antigen does type O blood have?
   A) A   C) A and B
   B) B   D) no antigen

10. What is the largest filtering lymph organ?
    A) spleen   C) tonsil
    B) thymus   D) node

Thinking Critically

11. Identify the following as having oxygen-rich or carbon dioxide-filled blood: *aorta, coronary arteries, coronary veins, inferior vena cava, left atrium, left ventricle, right atrium, right ventricle,* and *superior vena cava.*

12. Compare and contrast the three types of blood vessels.

13. Explain how the lymphatic system works with the cardiovascular system.

14. Why is cancer of the blood cells or lymph nodes hard to control?

15. Arteries are distributed throughout the body, yet a pulse usually is taken at the neck or wrist. Explain why.

Developing Skills

16. **Comparing and Contrasting** Compare the life spans of the red blood cells, white blood cells, and platelets.

17. **Interpreting Data** Interpret the data listed in this table. Find the average heart rate of the three males and the three females and compare the two averages.

Gender and Heart Rate	
Sex	**Pulse/Minute**
Male 1	72
Male 2	64
Male 3	65
Female 1	67
Female 2	84
Female 3	74

18. **Sequencing** Describe the sequence of blood clotting from the wound to forming a scab.

19. **Comparing and Contrasting** Compare and contrast the functions of arteries, veins, and capillaries.

20. Concept Mapping Complete the events chain concept map showing how lymph moves in your body.

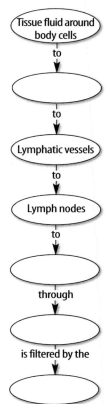

Tissue fluid around body cells

to

to

Lymphatic vessels

to

Lymph nodes

to

through

is filtered by the

Performance Assessment

21. Poster Prepare a poster illustrating heart transplants. Include an explanation of why the patient is given drugs that suppress the immune system and describe the patient's life after the operation.

22. Scientific Drawing Prepare a drawing of the human heart and label its parts.

TECHNOLOGY

Go to the Glencoe Science Web site at **science.glencoe.com** or use the **Glencoe Science CD-ROM** for additional chapter assessment.

THE PRINCETON REVIEW **Test Practice**

Ashley has investigated how different activities affect her body. She did five different activities for one minute each. After each activity, she recorded her pulse rate, body temperature, and degree of sweating as shown below.

Results from Ashley's Activities			
Activity	Pulse Rate (beats/min)	Body Temperature	Degree of Sweating
1	80	98.6°F	None
2	90	98.8°F	Minimal
3	100	98.9°F	Little
4	120	99.1°F	Moderate
5	150	99.5°F	Considerable

1. According to the information in this table, which activity caused Ashley's pulse to be greater than 120 beats per minute?
A) Activity 2
B) Activity 3
C) Activity 4
D) Activity 5

2. A reasonable hypothesis based on these data is that during Activity 1, Ashley was probably _____ .
F) walking quickly
G) sitting down
H) jogging
J) sprinting

Respiration and Excretion

How do you feel when you've just finished running a mile, or sliding into home base, or slamming a soccer ball into the goal past your opponent? If you're like most people, you probably breathe hard and perspire. Maybe you have even felt your lungs would burst. You need a constant supply of oxygen to keep your body cells functioning. Your body is adapted to meet that need.

What do you think?

Science Journal Look at the picture below. What do you think these wormlike things are? Here's a hint: *You are glad you have them on a dusty day.* Write your answer or best guess in your Science Journal.

EXPLORE ACTIVITY

Your body can store food and water, but it cannot store much oxygen. Breathing brings oxygen into your body. In the following activity, find out about one factor that can change your breathing rate.

Measure breathing rate

1. Put your hand on the side of your rib cage. Take a deep breath. Notice how your rib cage moves out and upward when you inhale.

2. Count the number of breaths you take for 15 s. Multiply this number by four to calculate your normal breathing rate for 1 min.

3. Repeat step 2 two more times, then calculate your average breathing rate.

4. Do a physical activity described by your teacher for 1 min and repeat step 2 to determine your breathing rate now.

5. Time how long it takes for your breathing rate to return to normal.

Observe

How does breathing rate appear to be related to physical activity? Write your answer in your Science Journal.

Before You Read

FOLDABLES
Reading & Study Skills

Making a Know-Want-Learn Study Fold Make the following Foldable to help identify what you already know and what you want to know about respiration.

1. Place a sheet of paper in front of you so the long side is at the top. Fold the paper in half from top to bottom.

2. Fold in both sides to divide the paper into thirds. Unfold the paper.

3. Cut through the top thickness of paper along each of the fold lines to the top fold to form three tabs. Label each tab as shown.

4. Before you read the chapter, write *I breathe* under the left tab. Write *Why do I breathe?* under the middle tab.

5. As you read the chapter, write the answer you learn under the right tab.

1 The Respiratory System

As You Read

What You'll Learn

- **Describe** the functions of the respiratory system.
- **Explain** how oxygen and carbon dioxide are exchanged in the lungs and in tissues.
- **Identify** the pathway of air in and out of the lungs.
- **Explain** the effects of smoking on the respiratory system.

Vocabulary

pharynx   alveoli
larynx   diaphragm
trachea   emphysema
bronchi   asthma

Why It's Important

Your body's cells depend on your respiratory system to supply oxygen and remove carbon dioxide.

Functions of the Respiratory System

Can you imagine an astronaut walking on the Moon without a space suit or a diver exploring the ocean without scuba gear? Of course not. You couldn't survive in either location under those conditions because you need to breathe air. Earth is surrounded by a layer of gases called the atmosphere (AT muh sfihr). You breathe atmospheric gases that are closest to Earth. As shown in **Figure 1,** oxygen is one of those gases.

For thousands of years people have known that air, food, and water are needed for life. However, the gas in the air that is necessary for life was not identified as oxygen until the late 1700s. At that time, a French scientist experimented and discovered that an animal breathed in oxygen and breathed out carbon dioxide. He measured the amount of oxygen that the animal used and the amount of carbon dioxide produced by its bodily processes. After his work with animals, the French scientist used this knowledge to study the way that humans use oxygen. He measured the amount of oxygen that a person uses when resting and when exercising. These measurements were compared, and he discovered that more oxygen is used by the body during exercise.

Figure 1
Air, which is needed by most organisms, is only 21 percent oxygen.

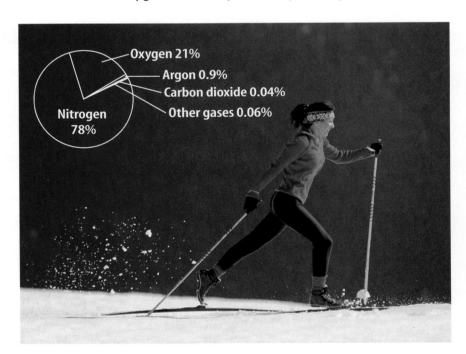

Oxygen 21%
Argon 0.9%
Carbon dioxide 0.04%
Other gases 0.06%
Nitrogen 78%

Figure 2
Several processes are involved in how the body obtains, transports, and uses oxygen.

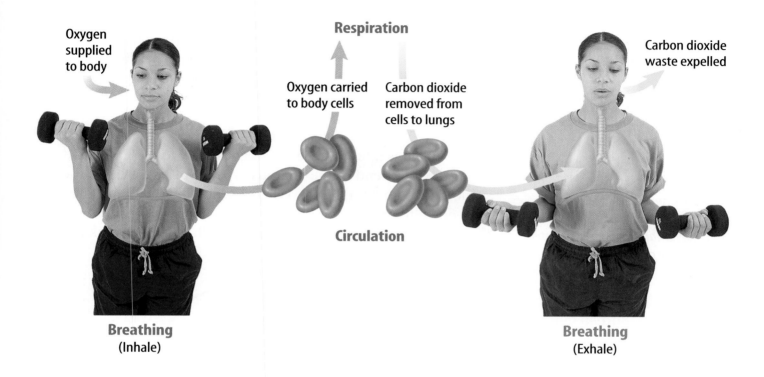

$$C_6H_{12}O_6 \ + \ 6O_2 \ \longrightarrow \ 6CO_2 \ + \ 6H_2O \ + \ \text{Energy}$$
Glucose + Oxygen ⟶ Carbon + Water + Energy
dioxide

Oxygen supplied to body

Respiration

Oxygen carried to body cells

Carbon dioxide removed from cells to lungs

Carbon dioxide waste expelled

Circulation

Breathing (Inhale)

Breathing (Exhale)

Breathing and Respiration People often confuse the terms *breathing* and *respiration*. Breathing is the movement of the chest that brings air into the lungs and removes waste gases. The air entering the lungs contains oxygen. It passes from the lungs into the circulatory system because there is less oxygen in the blood than in cells of the lungs. Blood carries oxygen to individual cells. At the same time, the digestive system supplies glucose from digested food to the same cells. The oxygen delivered to the cells is used to release energy from glucose. This chemical reaction, shown in the equation in **Figure 2,** is called cellular respiration. Without oxygen, this reaction would not take place. Carbon dioxide and water molecules are waste products of cellular respiration. They are carried back to the lungs in the blood. Exhaling, or breathing out, eliminates waste carbon dioxide and some water molecules.

Earth Science
INTEGRATION

The amount of water vapor in the atmosphere varies from almost none over deserts to nearly four percent in tropical rain forest areas. This means that every 100 molecules that make up air include only four molecules of water. In your Science Journal, infer how breathing dry air can stress your respiratory system.

 **Reading Check** *What is respiration?*

Organs of the Respiratory System

The respiratory system, shown in **Figure 3,** is made up of structures and organs that help move oxygen into the body and waste gases out of the body. Air enters your body through two openings in your nose called nostrils or through the mouth. Fine hairs inside the nostrils trap dust from the air. Air then passes through the nasal cavity, where it gets moistened and warmed by the body's heat. Glands that produce sticky mucus line the nasal cavity. The mucus traps dust, pollen, and other materials that were not trapped by nasal hairs. This process helps filter and clean the air you breathe. Tiny, hairlike structures, called cilia (SIHL ee uh), sweep mucus and trapped material to the back of the throat where it can be swallowed.

Pharynx Warmed, moist air then enters the **pharynx** (FER ingks), which is a tubelike passageway used by food, liquid, and air. At the lower end of the pharynx is a flap of tissue called the epiglottis (ep uh GLAHT us). When you swallow, your epiglottis folds down to prevent food or liquid from entering your airway. The food enters your esophagus instead. What do you think has happened if you begin to choke?

Figure 3
Air can enter the body through the nostrils and the mouth. *What is an advantage of having air enter through the nostrils?*

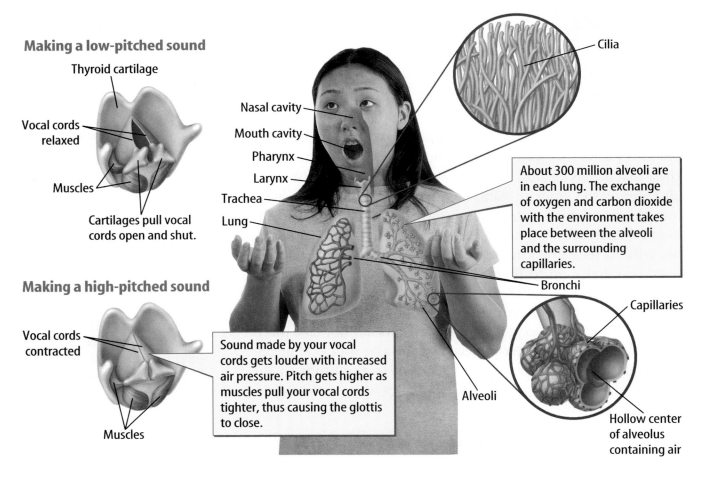

Making a low-pitched sound

Thyroid cartilage

Vocal cords relaxed

Muscles

Cartilages pull vocal cords open and shut.

Making a high-pitched sound

Vocal cords contracted

Muscles

Sound made by your vocal cords gets louder with increased air pressure. Pitch gets higher as muscles pull your vocal cords tighter, thus causing the glottis to close.

Cilia

Nasal cavity
Mouth cavity
Pharynx
Larynx
Trachea
Lung

About 300 million alveoli are in each lung. The exchange of oxygen and carbon dioxide with the environment takes place between the alveoli and the surrounding capillaries.

Bronchi

Capillaries

Alveoli

Hollow center of alveolus containing air

Larynx and Trachea Next, the air moves into your larynx (LER ingks). The **larynx** is the airway to which two pairs of horizontal folds of tissue, called vocal cords, are attached as shown in **Figure 3.** Forcing air between the cords causes them to vibrate and produce sounds. When you speak, muscles tighten or loosen your vocal cords, resulting in different sounds. Your brain coordinates the movement of the muscles in your throat, tongue, cheeks, and lips when you talk, sing, or just make noise. Your teeth also are involved in forming letter sounds and words.

From the larynx, air moves into the **trachea** (TRAY kee uh), which is a tube about 12 cm in length. Strong, C-shaped rings of cartilage prevent the trachea from collapsing. The trachea is lined with mucous membranes and cilia, as shown in **Figure 3,** that trap dust, bacteria, and pollen. Why must the trachea stay open all the time?

Bronchi and the Lungs Air is carried into your lungs by two short tubes called **bronchi** (BRAHN ki) (singular, *bronchus)* at the lower end of the trachea. Within the lungs, the bronchi branch into smaller and smaller tubes. The smallest tubes are called bronchioles (BRAHN kee ohlz). At the end of each bronchiole are clusters of tiny, thin-walled sacs called **alveoli** (al VEE uh li). Air passes into the bronchi, then into the bronchioles, and finally into the alveoli. As shown in **Figure 3,** lungs are masses of alveoli arranged in grapelike clusters. The capillaries surround the alveoli like a net.

The exchange of oxygen and carbon dioxide takes place between the alveoli and capillaries. This easily happens because the walls of the alveoli (singular, *alveolus)* and the walls of the capillaries are each only one cell thick, as shown in **Figure 4.** Oxygen moves through the cell membranes of the alveoli and then through the cell membranes of the capillaries into the blood. There the oxygen is picked up by hemoglobin (HEE muh gloh bun), a molecule in red blood cells, and carried to all body cells. At the same time, carbon dioxide and other cellular wastes leave the body cells. The wastes move through the cell membranes of the capillaries. Then they are carried by the blood. In the lungs, waste gases move through the cell membranes of the capillaries and through the cell membranes of the alveoli. Then waste gases leave the body during exhalation.

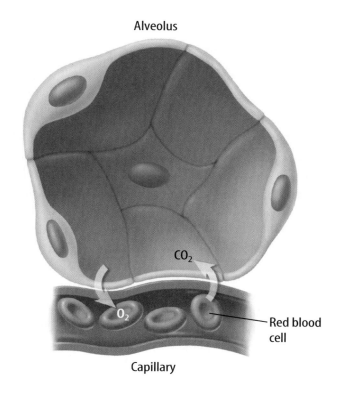

Alveolus

CO_2

O_2

Red blood cell

Capillary

Figure 4
The thin capillary walls allow gases to be exchanged easily between the alveoli and the capillaries.

SCIENCE *Online*

Research Visit the Glencoe Science Web site at **science.glencoe.com** for more information about how speech sounds are made. Report to your class what you learn.

TRY AT HOME Mini LAB

Comparing Surface Area

Procedure

1. Stand a **bathroom-tissue cardboard tube** in an **empty bowl.**
2. Drop **marbles** into the tube, filling it to the top.
3. Count the number of marbles used.
4. Repeat steps 2 and 3 two more times. Calculate the average number of marbles needed to fill the tube.
5. The tube's inside surface area is approximately 161.29 cm^2. Each marble has a surface area of approximately 8.06 cm^2. Calculate the surface area of the average number of marbles.

Analysis

1. Compare the inside surface area of the tube with the surface area of the average number of marbles needed to fill the tube.
2. If the tube represents a bronchus, what do the marbles represent?
3. Using this model, explain what makes gas exchange in the lungs efficient.

Why do you breathe?

Signals from your brain tell the muscles in your chest and abdomen to contract and relax. You don't have to think about breathing to breathe, just like your heart beats without you telling it to beat. Your brain can change your breathing rate depending on the amount of carbon dioxide present in your blood. If a lot of carbon dioxide is present, your breathing rate increases. It decreases if less carbon dioxide is in your blood. You do have some control over your breathing—you can hold your breath if you want to. Eventually, though, your brain will respond to the buildup of carbon dioxide in your blood. The brain's response will tell your chest and abdomen muscles to work automatically, and you will breathe whether you want to or not.

Inhaling and Exhaling Breathing is partly the result of changes in air pressure. Under normal conditions, a gas moves from an area of high pressure to an area of low pressure. When you squeeze an empty, soft-plastic bottle, air is pushed out. This happens because air pressure outside the top of the bottle is less than the pressure you create inside the bottle when you squeeze it. As you release your grip on the bottle, the air pressure inside the bottle becomes less than it is outside the bottle. Air rushes back in, and the bottle returns to its original shape.

Your lungs work in a similar way to the squeezed bottle. Your **diaphragm** (DI uh fram) is a muscle beneath your lungs that contracts and relaxes to help move gases into and out of your lungs. **Figure 5** illustrates breathing.

✓ **Reading Check** *How does your diaphragm help you breathe?*

When a person is choking, a rescuer can use abdominal thrusts, as shown in **Figure 6,** to save the life of the choking victim.

Figure 5
Your lungs inhale and exhale about 500 mL of air with an average breath. This increases to 2,000 mL of air per breath when you do strenuous activity.

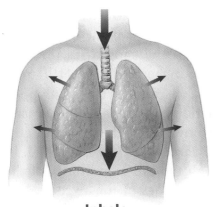
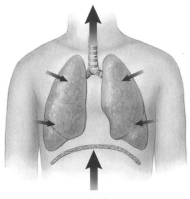

Inhale   Exhale

Figure 6

When food or other objects become lodged in the trachea, airflow between the lungs and the mouth and nasal cavity is blocked. Death can occur in minutes. However, prompt action by someone can save the life of a choking victim. The rescuer uses abdominal thrusts to force the victim's diaphragm up. This decreases the volume of the chest cavity and forces air up in the trachea. The result is a rush of air that dislodges and expels the food or other object. The victim can breathe again. This technique is shown at right and should only be performed in emergency situations.

Food is lodged in the victim's trachea.

The rescuer places her fist against the victim's stomach.

The rescuer's second hand adds force to the fist.

A The rescuer stands behind the choking victim and wraps her arms around the victim's upper abdomen. She places a fist (thumb side in) against the victim's stomach. The fist should be below the ribs and above the navel.

An upward thrust dislodges the food from the victim's trachea.

B With a violent, sharp movement, the rescuer thrusts her fist up into the area below the ribs. This action should be repeated as many times as necessary.

Table 1 Smokers' Risk of Death from Disease	
Disease	Smokers' Risk Compared to Nonsmokers' Risk
Lung Cancer	23 times higher for males, 11 times higher for females
Chronic Bronchitis and Emphysema	5 times higher
Heart Disease	2 times higher

Diseases and Disorders of the Respiratory System

Environmental Science
INTEGRATION

If you were asked to list some of the things that can harm your respiratory system, you probably would put smoking at the top. As you can see in **Table 1,** many serious diseases are related to smoking. The chemical substances in tobacco—nicotine and tars—are poisons and can destroy cells. The high temperatures, smoke, and carbon monoxide produced when tobacco burns also can injure a smoker's cells. Even if you are a nonsmoker, inhaling smoke from tobacco products—called secondhand smoke—is unhealthy and has the potential to harm your respiratory system. Smoking, polluted air, coal dust, and asbestos (as BES tus) have been related to respiratory problems such as bronchitis (brahn KITE us), emphysema (em fuh SEE muh), asthma (AZ muh), and cancer.

Respiratory Infections Bacteria, viruses, and other microorganisms can cause infections that affect any of the organs of the respiratory system. The common cold usually affects the upper part of the respiratory system—from the nose to the pharynx. The cold virus also can cause irritation and swelling in the larynx, trachea, and bronchi. The cilia that line the trachea and bronchi can be damaged. However, cilia usually heal rapidly. A virus that causes influenza, or flu, can affect many of the body's systems. The virus multiplies in the cells lining the alveoli and damages them. Pneumonia is an infection in the alveoli that can be caused by bacteria, viruses, or other microorganisms. Before antibiotics were available to treat these infections, many people died from pneumonia.

 **Reading Check** *What parts of the respiratory system are affected by the cold virus?*

SCIENCE *Online*

Research Visit the Glencoe Science Web site at **science.glencoe.com** for more information about the health aspects of second-hand smoke. Make a poster explaining what you learn.

Chronic Bronchitis When bronchial tubes are irritated and swell and too much mucus is produced, a disease called bronchitis develops. Sometimes, bacterial infections occur in the bronchial tubes because the mucus there provides nearly ideal conditions for bacteria to grow. Antibiotics are effective treatments for this type of bronchitis.

Many cases of bronchitis clear up within a few weeks, but the disease sometimes lasts for a long time. When this happens, it is called chronic (KRAHN ihk) bronchitis. A person who has chronic bronchitis must cough often to try to clear the excess mucus from the airway. However, the more a person coughs, the more the cilia and bronchial tubes can be harmed. When cilia are damaged, they cannot move mucus, bacteria, and dirt particles out of the lungs effectively. Then harmful substances, such as sticky tar from burning tobacco, build up in the airways. Sometimes, scar tissue forms and the respiratory system cannot function properly.

Emphysema A disease in which the alveoli in the lungs enlarge is called **emphysema** (em fuh SEE muh). When cells in the alveoli are reddened and swollen, an enzyme is released that causes the walls of the alveoli to break down. As a result, alveoli can't push air out of the lungs, so less oxygen moves into the bloodstream from the alveoli. When blood becomes low in oxygen and high in carbon dioxide, shortness of breath occurs. Some people with emphysema require extra oxygen as shown in **Figure 7C.** Because the heart works harder to supply oxygen to body cells, people who have emphysema often develop heart problems, as well.

Do you know what to do if someone is having trouble breathing? To find out about such emergencies, see the **Emergencies Field Guide** at the back of the book.

Figure 7
Lung diseases can have major effects on breathing.

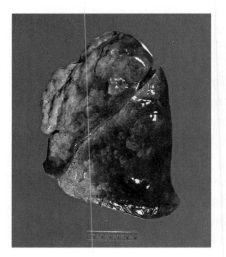

A A normal, healthy lung can exchange oxygen and carbon dioxide effectively.

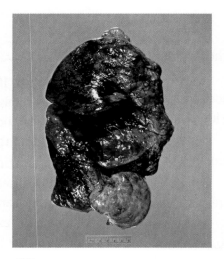

B A diseased lung carries less oxygen to body cells.

C Emphysema may take 20 to 30 years to develop.

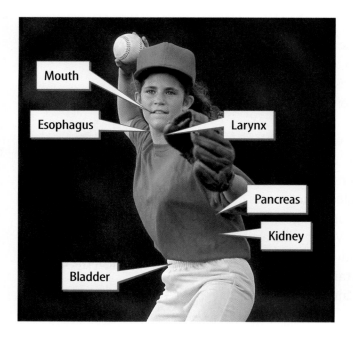

Mouth

Esophagus

Larynx

Pancreas

Kidney

Bladder

Lung Cancer The third leading cause of death in men and women in the United States is lung cancer. Inhaling the tar in cigarette smoke is the greatest contributing factor to lung cancer. Tar and other ingredients found in smoke act as carcinogens (kar SIHN uh junz) in the body. Carcinogens are substances that can cause an uncontrolled growth of cells. In the lungs, this is called lung cancer. Lung cancer is not easy to detect in its early stages. Smoking also has been linked to the development of cancers of the mouth, esophagus, larynx, pancreas, kidney, and bladder. See **Figure 8.**

✓ **Reading Check** *What do you think will happen to the lungs of a young person who begins smoking?*

Figure 8
More than 85 percent of all lung cancer is related to smoking. Smoking also can play a part in the development of cancer in other body organs indicated above.

Asthma Shortness of breath, wheezing, or coughing can occur in a lung disorder called **asthma.** When a person has an asthma attack, the bronchial tubes contract quickly. Inhaling medicine that relaxes the bronchial tubes is the usual treatment for an asthma attack. Asthma is often an allergic reaction. An allergic reaction occurs when the body overreacts to a foreign substance. An asthma attack can result from breathing certain substances such as cigarette smoke or certain plant pollen, eating certain foods, or stress in a person's life.

Section 1 Assessment

1. What is the main function of the respiratory system?

2. How are oxygen, carbon dioxide, and other waste gases exchanged in the lungs and body tissues?

3. What causes air to move into and out of a person's lungs?

4. How does smoking affect the respiratory and circulatory systems?

5. **Think Critically** How is the work of the digestive and circulatory systems related to the respiratory system?

Skill Builder Activities

6. **Researching Information** Nicotine in tobacco is a poison. Using library references, find out how nicotine affects the body. **For more help, refer to the** Science Skill Handbook.

7. **Communicating** Use references to find out about lung disease common among coal miners, stonecutters, and sandblasters. Find out what safety measures are required now for these trades. In your Science Journal, write a paragraph about these safety measures. **For more help, refer to the** Science Skill Handbook.

The Excretory System

Functions of the Excretory System

It's your turn to take out the trash. You carry the bag outside and put it in the trash can. The next day, you bring out another bag of trash, but the trash can is full. When trash isn't collected, it piles up. Just as trash needs to be removed from your home to keep it livable, your body must eliminate wastes to remain healthy. Undigested material is eliminated by your large intestine. Waste gases are eliminated through the combined efforts of your circulatory and respiratory systems. Some salts are eliminated when you sweat. These systems function together as parts of your excretory system. If wastes aren't eliminated, toxic substances build up and damage organs. If not corrected, serious illness or death occurs.

The Urinary System

The **urinary system** rids the blood of wastes produced by the cells. **Figure 9** shows how the urinary system functions as a part of the excretory system. The urinary system also controls blood volume by removing excess water produced by body cells during respiration.

As You Read

What **You'll Learn**

- **Distinguish** between the excretory and urinary systems.
- **Describe** how the kidneys work.
- **Explain** what happens when urinary organs don't work.

Vocabulary

urinary system	ureter
urine	bladder
kidney	urethra
nephron	

Why **It's Important**

The urinary system helps clean your blood of cellular wastes.

Figure 9
The urinary system, along with the digestive and respiratory systems, and the skin make up the excretory system.

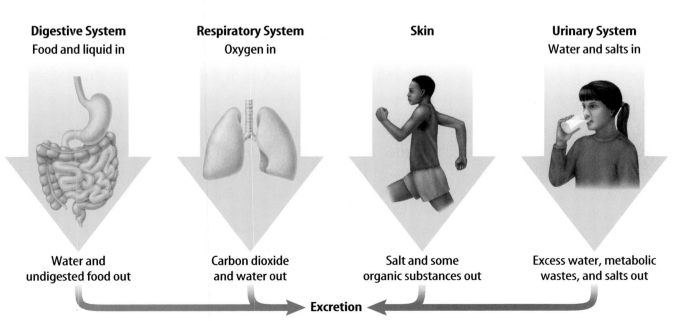

Digestive System
Food and liquid in

Respiratory System
Oxygen in

Skin

Urinary System
Water and salts in

Water and undigested food out

Carbon dioxide and water out

Salt and some organic substances out

Excess water, metabolic wastes, and salts out

Excretion

Figure 10
The amount of urine that you eliminate each day is determined by the level of a hormone that is produced by your hypothalamus.

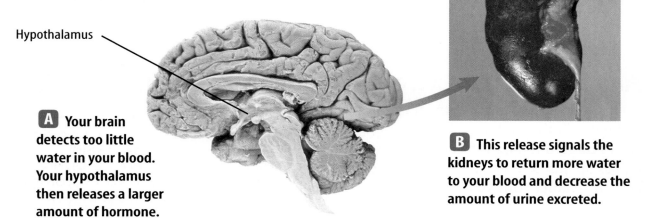

Hypothalamus

A Your brain detects too little water in your blood. Your hypothalamus then releases a larger amount of hormone.

B This release signals the kidneys to return more water to your blood and decrease the amount of urine excreted.

Regulating Fluid Levels To stay in good health, the fluid levels within the body must be balanced and normal blood pressure must be maintained. An area in the brain, the hypothalamus (hi poh THAL uh mus), constantly monitors the amount of water in the blood. When the brain detects too much water in the blood, the hypothalamus releases a lesser amount of a specific hormone. This signals the kidneys to return less water to the blood and increase the amount of wastewater, called **urine,** that is excreted. **Figure 10** shows what happens when too little water is in the blood.

Reading Check *How does the urinary system control the volume of water in the blood?*

A specific amount of water in the blood is also important for the movement of gases and excretion of solid wastes from the body. The urinary system also balances the amounts of certain salts and water that must be present for all cell activities to take place.

Organs of the Urinary System Excretory organs is another name for the organs of the urinary system. The main organs of the urinary system are two bean-shaped **kidneys.** Kidneys are located on the back wall of the abdomen at about waist level. The kidneys filter blood that contains wastes collected from cells. In approximately 5 min, all of the blood in your body passes through the kidneys. The red-brown color of the kidneys is due to their enormous blood supply. In **Figure 11A,** you can see that blood enters the kidneys through a large artery and leaves through a large vein.

Filtration in the Kidney The kidney, shown in **Figure 11B,** is a two-stage filtration system. It is made up of about 1 million tiny filtering units called **nephrons** (NEF rahnz), shown in **Figure 11C.** Each nephron has a cuplike structure and a tubelike structure called a duct. Blood moves from a renal artery to capillaries in the cuplike structure. The first filtration occurs when water, sugar, salt, and wastes from the blood pass into the cuplike structure. Left behind in the blood are the red blood cells and proteins. Next, liquid in the cuplike structure is squeezed into a narrow tubule. Capillaries that surround the tubule perform the second filtration. Most of the water, sugar, and salt are reabsorbed and returned to the blood. These collection capillaries merge to form small veins, which merge to form a renal vein in each kidney. Purified blood is returned to the main circulatory system. The liquid left behind flows into collecting tubules in each kidney. This wastewater, or urine, contains excess water, salts, and other wastes that are not reabsorbed by the body. An average-sized person produces about 1 L of urine per day.

Figure 11
The urinary system removes wastes from the blood.

A **The urinary system includes the kidneys, the bladder, and the connecting tubes.**

B **Kidneys are made up of many nephrons.**

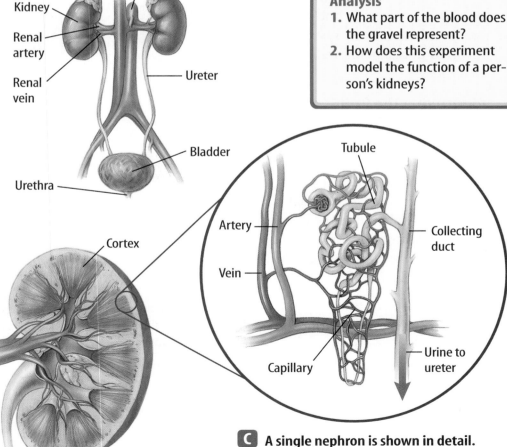

C **A single nephron is shown in detail.**
What is the main function of the nephron?

Mini LAB

Modeling Kidney Function

Procedure

1. Mix a small amount of **soil** and **fine gravel** with **water** in a **clean cup.**
2. Place the **funnel** into a **second cup.**
3. Place a small piece of **wire screen** in the funnel.
4. Carefully pour the mud-water-gravel mixture into the funnel. Let it drain.
5. Remove the screen and replace it with a piece of **filter paper.**
6. Place the funnel in **another clean cup.**
7. Repeat step 4.

Analysis

1. What part of the blood does the gravel represent?
2. How does this experiment model the function of a person's kidneys?

Urine Collection and Release The urine in each collecting tubule drains into a funnel-shaped area of each kidney that leads to the ureter (YER ut ur). **Ureters** are tubes that lead from each kidney to the bladder. The **bladder** is an elastic, muscular organ that holds urine until it leaves the body. The elastic walls of the bladder can stretch to hold up to 0.5 L of urine. When empty, the bladder looks wrinkled and the cells lining the bladder are thick. When full, the bladder looks like an inflated balloon and the cells lining the bladder are stretched and thin. A tube called the **urethra** (yoo REE thruh) carries urine from the bladder to the outside of the body.

Problem-Solving Activity

How does your body gain and lose water?

Your body depends on water. Without water, your cells could not carry out their activities and body systems could not function. Water is so important to your body that your brain and other body systems are involved in balancing water gain and water loss.

Identifying the Problem

Table A shows the major sources by which your body gains water. Oxidation of nutrients occurs when energy is released from nutrients by your body's cells. Water is a waste product of these reactions. **Table B** lists the major sources by which your body loses water. The data show you how daily gain and loss of water are related.

Solving the Problem

1. What is the greatest source of water gained by your body?
2. How would the percentages of water gained and lost change in a person who was working in extremely warm temperatures? In this case, what organ of the body would be the greatest contributor to water loss?

Table A

Major Sources by Which Body Water Is Gained		
Source	Amount (mL)	Percent
Oxidation of Nutrients	250	10
Foods	750	30
Liquids	1,500	60
Total	**2,500**	**100**

Table B

Major Sources by Which Body Water Is Lost		
Source	Amount (mL)	Percent
Urine	1,500	60
Skin	500	20
Lungs	350	14
Feces	150	6
Total	**2,500**	**100**

Other Organs of Excretion

Large amounts of liquid wastes are lost every day by your body in other ways, as shown in **Figure 12**. The liver also filters the blood to remove wastes. Certain wastes are converted to other substances. For example, excess amino acids are changed to urea (yoo REE uh), which is a chemical that ends up in urine. Hemoglobin from broken-down red blood cells becomes part of bile, which is the digestive fluid from the liver.

Urinary Diseases and Disorders

What happens when someone's kidneys don't work properly or stop working? Waste products that are not removed build up and act as poisons in body cells. Water that normally is removed from body tissues accumulates and causes swelling of the ankles and feet. Sometimes these fluids also build up around the heart, and it has to work harder to move blood to the lungs.

Without excretion, an imbalance of salts occurs. The body responds by trying to restore this balance. If the balance isn't restored, the kidneys and other organs can be damaged. Kidney failure occurs when the kidneys don't work as they should. This is always a serious problem because the kidneys' job is so important to the rest of the body.

Infections caused by microorganisms can affect the urinary system. Usually, the infection begins in the bladder. However, it can spread and involve the kidneys. Most of the time, these infections can be cured with antibiotics.

Because the ureters and urethra are narrow tubes, they can be blocked easily in some disorders. A blockage of one of these tubes can cause serious problems because urine cannot flow out of the body properly. If the blockage is not corrected, the kidneys can be damaged.

> **✔ Reading Check** *Why is a blocked ureter or urethra a serious problem?*

Detecting Urinary Diseases Urine can be tested for any signs of a urinary tract disease. A change in the urine's color can suggest kidney or liver problems. High levels of glucose can be a sign of diabetes. Increased amounts of a protein called albumin (al BYEW mun) indicate kidney disease or heart failure. When the kidneys are damaged, albumin can get into the urine, just as a leaky water pipe allows water to drip.

Figure 12
On average, the volume of water lost daily by exhaling is a little more than the volume of a soft-drink can. The volume of water lost by your skin each day is about the volume of a 20-ounce soft-drink bottle.

Earth Science
INTEGRATION

Nearly 80 percent of Earth's surface is covered by water. Ninety-seven percent of this water is salt water. Humans cannot drink salt water, so they depend on the less than one percent of freshwater that is available for use. In your Science Journal, infer how your kidneys would need to be different for you to be able to drink salt water.

Figure 13

A dialysis machine can replace or help with some of the activities of the kidneys in a person with kidney failure. Like the kidney, the dialysis machine removes wastes from the blood.

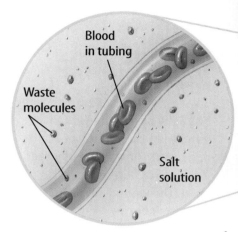

Waste molecules

Blood in tubing

Salt solution

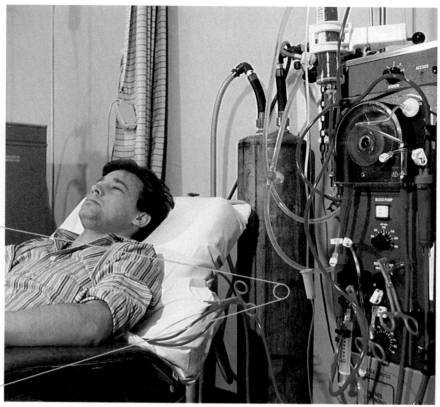

Dialysis A person who has only one kidney still can live normally. The remaining kidney increases in size and works harder to make up for the loss of the other kidney. However, if both kidneys fail, the person will need to have his or her blood filtered by an artificial kidney machine in a process called dialysis (di AL uh sus), as shown in **Figure 13.**

Section 2 Assessment

1. Describe the functions of a person's urinary system.

2. Explain how the kidneys remove wastes and keep fluids and salts in balance.

3. Describe what happens when the urinary system does not function properly.

4. Compare the excretory system and urinary system.

5. **Think Critically** Explain why reabsorption of certain materials in the kidneys is important to your health.

Skill Builder Activities

6. **Concept Mapping** Using a network tree concept map, compare the excretory functions of the kidneys and the lungs. **For more help, refer to the** Science Skill Handbook.

7. **Solving One-Step Equations** In approximately 5 min, all 5 L of blood in the body pass through the kidneys. Calculate the average rate of flow through the kidneys in liters per minute. **For more help, refer to the** Math Skill Handbook.

Activity

Kidney Structure

As your body uses nutrients, wastes are created. One role of the kidneys is to filter waste products out of the bloodstream and excrete this waste outside the body. How can these small structures filter all the blood in the body in 5 min?

What You'll Investigate
How does the structure of the kidney relate to the function of a kidney?

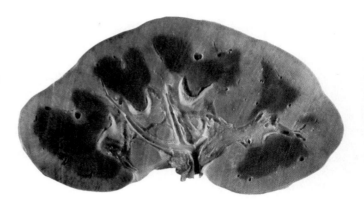

Materials
large animal kidney
model of a kidney
scalpel
hand lens
disposable gloves
Alternate material

Goal
■ **Observe** the external and internal structures of a kidney.

Safety Precautions

WARNING: *Use extreme care when using sharp instruments. Wear disposable gloves. Wash your hands with soap after completing this activity.*

Procedure

1. **Examine** the outside of the kidney supplied by your teacher.
2. If the kidney still is encased in fat, peel off the fat carefully.
3. Using a scalpel, carefully cut the tissue in half lengthwise around the outline of the kidney. This cut should result in a section similar to the illustration on this page.
4. **Observe** the internal features of the kidney using a hand lens, or view these features in a model.
5. **Compare** the specimen or model with the kidney in the illustration.
6. **Draw** the kidney in your Science Journal and label its structures.

Conclude and Apply

1. What part makes up the cortex of the kidney? Why is this part red?
2. What is the main function of nephrons?
3. The medulla of the kidney is made up of a network of tubules that come together to form the ureter. What is the function of this network of tubules?
4. How can the kidney be compared to a portable water-purifying system?

*C*ommunicating Your Data

Compare your conclusions with those of other students in your class. **For more help, refer to the** Science Skill Handbook.

Activity
Model and Invent

Simulating the Abdominal Thrust Maneuver

Have you ever taken a class in CPR or learned about how to help a choking victim? Using the abdominal thrust maneuver, or Heimlich maneuver, is one way to remove food or another object that is blocking someone's airway. What happens internally when the maneuver is used? How can you simulate the internal effects of the abdominal thrust maneuver?

Recognize the Problem

How can you simulate the removal of an object from the trachea when the abdominal thrust maneuver is used?

Thinking Critically

What can you use to make a model of the trachea? How can you simulate what happens during an abdominal thrust maneuver using your model?

Goals

- **Construct** a model of the trachea with a piece of food stuck in it.
- **Demonstrate** what happens when the abdominal thrust maneuver is performed on someone.

- **Predict** another way that air could get into the lungs if the food could not be dislodged with an abdominal thrust maneuver.

Possible Materials

paper towel roll or other tube
paper (wadded into a ball)
clay
bicycle pump
sports bottle
scissors

Safety Precautions

Always be careful when you use scissors.

Planning the Model

1. **List** the materials that you will need to construct your model. What will represent the trachea and a piece of food or other object blocking the airway?

2. How can you use your model to simulate the effects of an abdominal thrust maneuver?

3. Suggest a way to get air into the lungs if the food could not be dislodged. How would you simulate this method in your model?

Check the Model Plans

1. **Compare** your plans for the model and the abdominal thrust maneuver simulation with those of other students in your class. Discuss why each of you chose the plans and materials that you did.

2. Make sure your teacher approves your plan and materials for your model before you start.

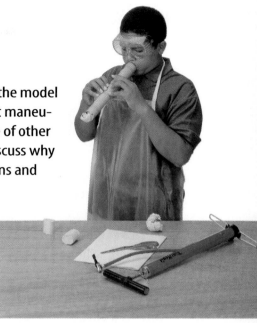

Making the Model

1. **Construct** your model of a trachea with an object stuck in it. Make sure that air cannot get through the trachea if you try blowing softly through it.

2. Simulate what happens when an abdominal thrust maneuver is used. Record your observations. Was the object dislodged? How hard was it to dislodge the object?

3. Replace the object in the trachea. Use your model to simulate how you could get air into the lungs if an abdominal thrust maneuver did not remove the object. Is it easy to blow air through your model now?

4. Model a crushed trachea. Is it easy to blow air through the trachea in this case?

Analyzing and Applying Results

1. **Describe** how easy it was to get air through the trachea in each step in the Making the Model section above. Include any other observations that you made as you worked with your model.

2. Think about what you did to get air into the trachea when the object could not be dislodged with an abdominal thrust maneuver. How could this be done to a person? Do you know what this procedure is called?

3. **Explain** why the trachea has cartilage around it to protect it. What might happen if it did not?

Explain to your family or friends what you have learned about how the abdominal thrust maneuver can help choking victims.

Overcoming the Odds

Guts and determination helped one pioneering doctor to save the lives of thousands

Overcoming the odds—especially when the odds seem stacked against you—is a challenge that many people face. Dr. Samuel Lee Kountz, Jr. (photo, right) had the odds stacked against him. Thanks to his determination he beat them.

Samuel Kountz decided at age eight to become a doctor. He faced his first challenge when he failed the entrance exam to his local Arkansas college. That didn't stop him, though. He asked the college president to give him another chance, and the president did. Kountz got into school and earned As and Bs. Kountz went on to get a graduate degree in biochemistry and was admitted to the University of Arkansas's medical school. For many, these achievements would be more than enough. But for Dr. Kountz, it was just the beginning of his quest to improve medicine—and to change history.

Dr. Kountz was especially interested in a process that was still brand new in the 1950s—the kidney transplant. For many patients, a kidney transplant added months or a year to one's life. But then a patient's body would reject the kidney, and the patient would die. Dr. Kountz was determined to see that kidney transplants saved lives and kept patients healthy for years.

Fixing the Problem

Kountz discovered the root of the problem—why and how a patient's body rejected the transplanted kidney. He discovered that the patient's cells attacked and destroyed the small blood vessels of the transplanted kidney. So the new kidney would die from lack of blood-supplied oxygen. He and others at Stanford University developed a way for doctors to watch the flow of the kidney's blood supply following surgery. Then doctors can give patients the right kinds of drugs at the right time, so that their bodies can overcome the rejection process.

In 1959, Kountz performed the first successful kidney transplant. He went on to develop a procedure to keep body organs healthy for up to 60 hours after being taken from a donor. He also set up a system of organ donor cards through the National Kidney Foundation. And in his career, Dr. Kountz transplanted more than 1,000 kidneys himself—and paved the way for thousands more.

A donated organ is on its way to save a life.

HUMAN ORGANS

UP

B 063 01

CONNECTIONS Research What kinds of medical breakthroughs has the last century brought? Locate an article that explains either a recent advance in medicine or the work that doctors and medical researchers are doing. Share your findings with your class.

SCIENCE Online

For more information, visit science.glencoe.com

Reviewing Main Ideas

Section 1 The Respiratory System

1. The respiratory system brings oxygen into the body and removes carbon dioxide.

2. Inhaled air passes through the nasal cavity, pharynx, larynx, trachea, bronchi, and into the alveoli of the lungs. *Why does the trachea, shown in the illustration, have cartilage but the esophagus does not?*

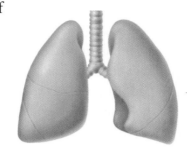

3. Breathing is the movement of the chest that brings air into the lungs and removes waste gases. The chemical reaction in the cells that needs oxygen to release energy from glucose is called cellular respiration.

4. The exchange of oxygen and carbon dioxide happens by the process of diffusion. In the lungs, oxygen diffuses into the capillaries from the alveoli. Carbon dioxide diffuses from the capillaries into the alveoli. In the body tissues, oxygen diffuses from the capillaries into the cells. Carbon dioxide diffuses from the cells into the capillaries.

5. Smoking causes many problems throughout the respiratory system, including chronic bronchitis, emphysema, and lung cancer. *What other body system is affected severely by smoking?*

Section 2 The Excretory System

1. The kidneys are the major organs of the urinary system. They filter wastes from all of the blood in the body. *How do the kidneys, shown in the photo regulate fluid levels in the body?*

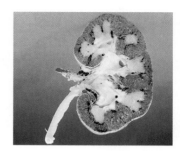

2. The kidney is a two-stage filtration system. The first filtration occurs when water, sugar, salt, and wastes from the blood pass into the cuplike part of the nephron. The capillaries surrounding the tubule part of the nephron perform the second filtration. In this filtration, most of the water, sugar, and salt are reabsorbed and returned to the blood.

3. The urinary system is part of the excretory system. The skin, lungs, liver, and large intestine are also excretory organs.

4. Urine can be tested for signs of urinary tract disease and other diseases.

5. A person who has only one kidney still can live normally. When kidneys fail to work, an artificial kidney can be used to filter the blood in a process called dialysis.

FOLDABLES
Reading & Study Skills

After You Read

Now that you've read the chapter, write the answer to the *Want* question under the *Learned* tab of your Foldable.

Visualizing Main Ideas

Complete the following table on the respiratory and excretory systems.

Human Body Systems	Respiratory System	Excretory System
Major Organs		
Wastes Eliminated		
Disorders		

Vocabulary Review

Vocabulary Words

a. alveoli
b. asthma
c. bladder
d. bronchi
e. diaphragm
f. emphysema
g. kidney
h. larynx
i. nephron
j. pharynx
k. trachea
l. ureter
m. urethra
n. urinary system
o. urine

 **THE PRINCETON REVIEW** **Study Tip**

Listening is a learning tool, too. Try recording a reading of your notes on tape and replaying it for yourself a few times a week.

Using Vocabulary

For each set of vocabulary words below, explain the relationship that exists.

1. alveoli, bronchi
2. bladder, urine
3. larynx, pharynx
4. ureter, urethra
5. alveoli, emphysema
6. nephron, kidney
7. urethra, bladder
8. asthma, bronchi
9. kidney, urine
10. diaphragm, alveoli

Chapter 4 Assessment

Checking Concepts

Choose the word or phrase that best answers the question.

1. When you inhale, which of the following contracts and moves down?
   A) bronchioles   C) nephrons
   B) diaphragm   D) kidneys

2. Air is moistened, filtered, and warmed in which of the following structures?
   A) larynx   C) nasal cavity
   B) pharynx   D) trachea

3. Exchange of gases occurs between capillaries and which of the following structures?
   A) alveoli   C) bronchioles
   B) bronchi   D) trachea

4. Which of the following is a lung disorder that can occur as an allergic reaction?
   A) asthma   C) emphysema
   B) atherosclerosis D) cancer

5. When you exhale, which way does the rib cage move?
   A) up   C) out
   B) down   D) stays the same

6. Which of the following conditions does smoking worsen?
   A) arthritis   C) excretion
   B) respiration   D) emphysema

7. Urine is held temporarily in which of the following structures?
   A) kidneys   C) ureter
   B) bladder   D) urethra

8. What are the filtering units of the kidneys?
   A) nephrons   C) neurons
   B) ureters   D) alveoli

9. Approximately 1 L of water is lost per day through which of the following?
   A) sweat   C) urine
   B) lungs   D) large intestine

10. Which of the following substances is not reabsorbed by blood after it passes through the kidneys?
    A) salt   C) wastes
    B) sugar   D) water

Thinking Critically

11. Explain why certain foods, such as peanuts, can cause choking in small children.

12. Why is it an advantage to have lungs with many smaller air sacs instead of having just two large sacs, like balloons?

13. Explain the damage to cilia, alveoli, and lungs from smoking.

14. What happens to the blood if the kidneys stop working?

15. Small, solid particles called kidney stones can form in the kidneys. Explain why it is often painful when a kidney stone passes into the ureter.

Developing Skills

16. **Interpreting Data** Study the data below. How much of each substance is reabsorbed into the blood in the kidneys? What substance is excreted completely in the urine?

Materials Filtered by the Kidneys		
Substance Filtered in Urine	Amount Moving Through Kidney	Amount Excreted
Water	125 L	1 L
Salt	350 g	10 g
Urea	1 g	1 g
Glucose	50 g	0 g

17. **Recognizing Cause and Effect** Discuss how lack of oxygen is related to lack of energy.

18. **Making and Using Graphs** Make a circle graph of total lung capacity using the following data:
    - volume of air in a normal inhalation or exhalation = 500 mL
    - volume of additional air that can be inhaled forcefully after a normal inhalation = 3,000 mL
    - volume of additional air that can be exhaled forcefully after a normal expiration = 1,100 mL
    - volume of air still left in the lungs after all the air that can be exhaled has been forcefully exhaled = 1,200 mL

19. **Forming Hypotheses** Make a hypothesis about the number of breaths a person might take per minute in each of these situations: asleep, exercising, and on top of Mount Everest. Give a reason for each hypothesis.

20. **Concept Mapping** Make an events chain concept map showing how urine forms in the kidneys. Begin with, "In the nephron …"

Performance Assessment

21. **Questionnaire and Interview** Prepare a questionnaire that can be used to interview a health specialist who works with lung cancer patients. Include questions on reasons for choosing the career, new methods of treatment, and the most encouraging or discouraging part of the job.

TECHNOLOGY

Go to the Glencoe Science Web site at **science.glencoe.com** or use the **Glencoe Science CD-ROM** for additional chapter assessment.

Test Practice

For one week, research scientists collected and accurately measured the amount of body water lost and gained per day for four different patients. They placed their results in the following table.

Body Water Gained (+) and Lost (−)				
Person	**Day 1** (L)	**Day 2** (L)	**Day 3** (L)	**Day 4** (L)
Mr. Stoler	+ 0.05	+ 0.15	− 0.35	+ 0.12
Mr. Jemma	− 0.01	0.00	− 0.20	− 0.01
Mr. Lowe	0.00	+ 0.10	− 0.28	+ 0.01
Mr. Cheng	− 0.50	− 0.50	− 0.55	− 0.32

Study the table and answer the following questions.

1. According to this information, which patient may be suffering from dehydration or an excessive amount of body water loss?
   A) Mr. Stoler
   B) Ms. Jemma
   C) Mr. Lowe
   D) Mr. Cheng

2. According to the table, it was probably very hot in each patient's hospital room during _____ .
   F) day one
   G) day two
   H) day three
   J) day four

Control and Coordination

O ne second, the puck is halfway across the ice. In the next second you're trying to stop it from making a goal. For a hockey goalie, keen eyesight is not enough. He needs to be able to respond quickly, without even thinking about it. In this chapter, you will learn how your body senses and responds to the world around you.

What do you think?

Science Journal Look at the picture below with a classmate. Discuss what you think this must be. Here's a hint: *It can separate the bitter from the sweet.* Write your answer or best guess in your Science Journal.

 **EXPLORE ACTIVITY**

If the weather is cool, you might put on a jacket. If you see friends, you might call out to them. You also might pick up a crying baby. Every second of the day you react to different sights, sounds, and smells in your environment. You control some of these reactions, but others take place in your body without thought. Some reactions protect you from harm. Do the activity below to see how one response can keep your body safe.

Observe a response

1. Wearing safety goggles, sit on a chair 1 m away from a partner.
2. Ask your partner to toss a wadded-up piece of paper at your face without warning you.
3. Switch positions and repeat the activity.

Observe

Describe in your Science Journal how you reacted to the ball of paper being thrown at you. Explain how your anticipation of being hit altered your body's response.

Before You Read

FOLDABLES
Reading & Study Skills

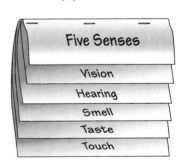

Making a Main Ideas Study Fold A main idea consists of the major concepts or topics talked about in a chapter. Before you read the chapter, make the following Foldable to help you identify the main idea(s) of this chapter.

1. Stack three sheets of paper in front of you so the short side of all sheets is at the top.
2. Slide the top sheet up so that about 4 cm of the middle sheet show. Slide the bottom sheet down so that about 4 cm of it shows below the middle sheet.
3. Fold the sheets top to bottom to form six tabs and staple along the top fold, as shown.
4. Label the flaps *Five Senses*, *Vision*, *Hearing*, *Smell*, *Taste*, and *Touch*, as shown. Before you read the chapter, write what you know about the five senses under the tabs.
5. As you read the chapter, add to or change the information you wrote under the tabs.

The Nervous System

As You Read

What You'll Learn

- **Describe** the basic structure of a neuron and how an impulse moves across a synapse.
- **Compare** the central and peripheral nervous systems.
- **Explain** how drugs affect the body.

Vocabulary

homeostasis
neuron
dendrite
axon
synapse
central nervous system
peripheral nervous system
cerebrum
cerebellum
brain stem
reflex

Why It's Important

Your body reacts to your environment because of your nervous system.

How the Nervous System Works

After doing the dishes and finishing your homework, you settle down in your favorite chair and pick up that mystery novel you've been trying to finish. Only three pages to go . . . Who did it? Why did she do it? Crash! You scream. What made that unearthly noise? You turn around to find that your dog's wagging tail has just swept the lamp off the table. Suddenly, you're aware that your heart is racing and your hands are shaking. After a few minutes though, your breathing returns to normal and your heartbeat is back to its regular rate. What's going on?

Responding to Stimuli The scene described above is an example of how your body responds to changes in its environment. Any internal or external change that brings about a response is called a stimulus (STIHM yuh lus). Each day, you're bombarded by thousands of stimuli, as shown in **Figure 1.** Noise, light, the smell of food, and the temperature of the air are all stimuli from outside your body. Chemical substances such as hormones are examples of stimuli from inside your body. Your body adjusts to changing stimuli with the help of your nervous system.

Figure 1
Stimuli are found everywhere and all the time, even when you're enjoying being with your friends. *What types of stimuli are present at this party?*

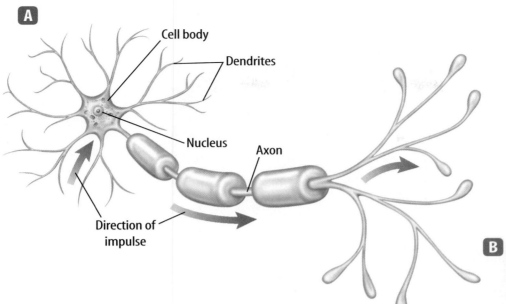

A
Cell body
Dendrites
Nucleus
Axon
Direction of impulse

Figure 2
A A neuron is made up of a cell body, dendrites, and an axon. *How does the branching of the dendrites allow for more impulses to be picked up by the neuron?* **B** *In this photograph of brain neurons, can you find the cell parts named in* **A** *?*

B

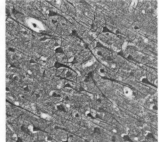

Homeostasis It's amazing how your body handles all these stimuli. Control systems maintain steady internal conditions. The regulation of steady, life-maintaining conditions inside an organism, despite changes in its environment, is called **homeostasis.** Examples of homeostasis are the regulation of your breathing, heartbeat, and digestion. Your nervous system is one of several control systems used by your body to maintain homeostasis.

Nerve Cells

The basic functioning units of the nervous system are nerve cells, or **neurons** (NOO rahnz). As shown in **Figure 2,** a neuron is made up of a cell body and branches called dendrites and axons. **Dendrites** receive messages from other neurons and send them to the cell body. **Axons** (AK sahns) carry messages away from the cell body. Any message carried by a neuron is called an impulse. Notice the branching at the end of the axon. This allows the impulses to move to many other muscles, neurons, or glands.

Types of Nerve Cells Your body has sensory receptors that produce electrical impulses and respond to stimuli, such as changes in temperature, sound, pressure, and taste. Three types of neurons—sensory neurons, motor neurons, and interneurons—transport impulses. Sensory neurons receive information and send impulses to the brain or spinal cord, where interneurons relay these impulses to motor neurons. Motor neurons then conduct impulses from the brain or spinal cord to muscles or glands throughout your body.

Field
GUIDE

Would you know what to do if you saw someone unconscious? To find out what to do in a situation such as this, see the **Emergencies Field Guide** at the back of the book.

Figure 3

Millions of nerve impulses are moving throughout your body as you read this page. In response to stimuli, many impulses follow a specific pathway —from sensory neuron to interneuron to motor neuron— to bring about a response. Like a relay team, these three types of neurons work together. The illustration on this page shows how the sound of a breaking window might startle you and cause you to drop a glass of water.

SENSORY NEURONS When you hear a loud noise, receptors in your ears—the specialized endings of sensory neurons—are stimulated. These sensory neurons produce nerve impulses that travel to your brain.

INTERNEURONS Interneurons in your brain receive the impulses from sensory neurons and pass them along to motor neurons.

MOTOR NEURONS Impulses travel down the axons of motor neurons to muscles—in this case, your biceps— which contract to jerk your arms in response to the loud noise.

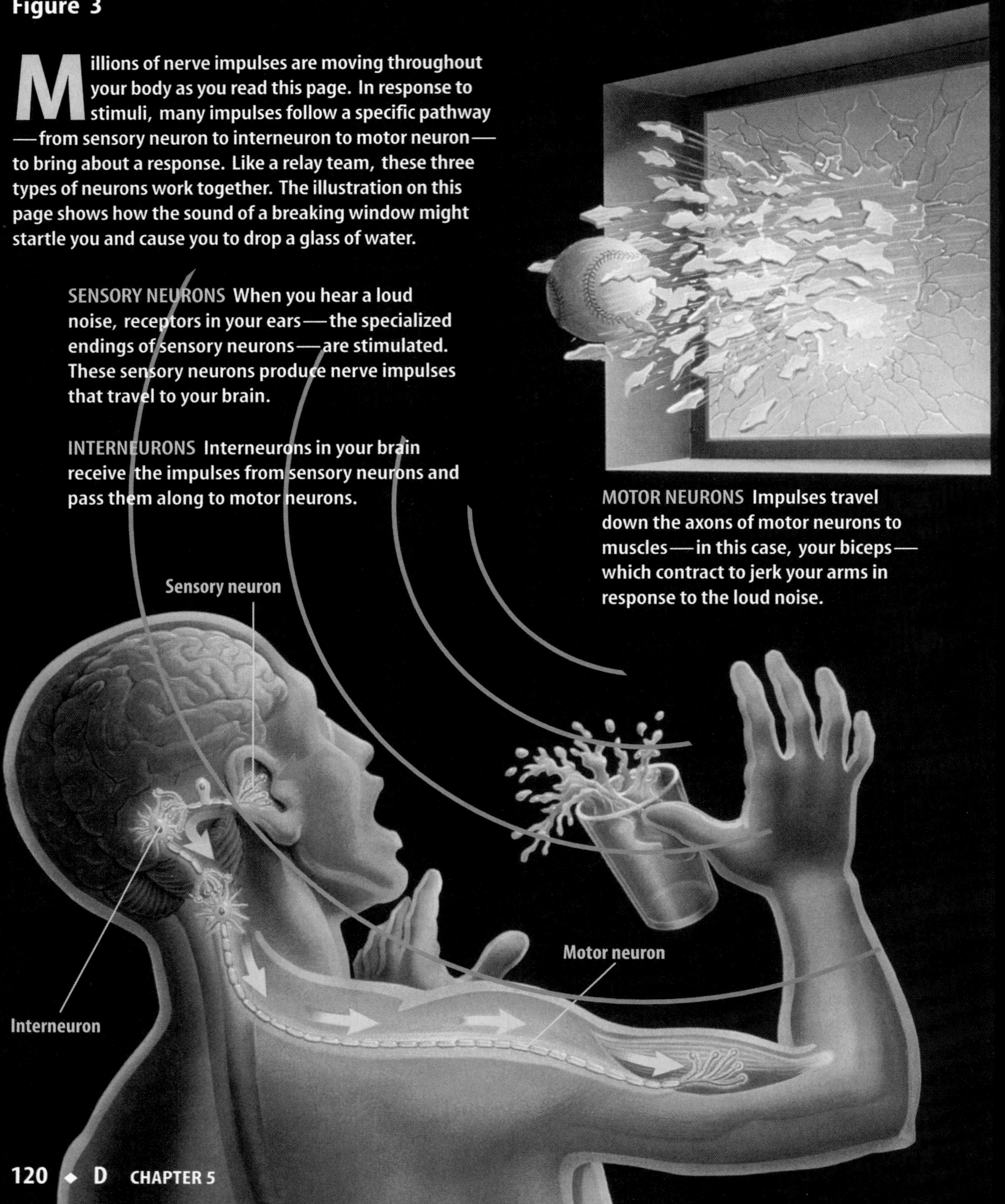

Sensory neuron

Interneuron

Motor neuron

Figure 4
An impulse moves in only one direction across a synapse—from an axon to the dendrites or cell body of another neuron.

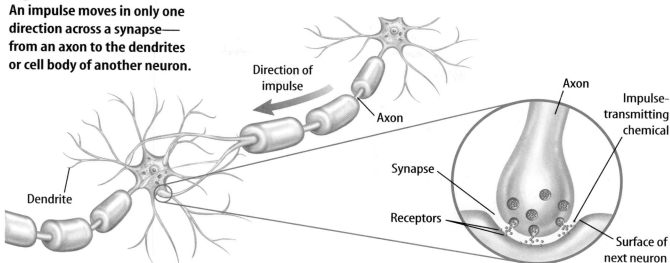

Direction of impulse

Axon

Dendrite

Axon

Impulse-transmitting chemical

Synapse

Receptors

Surface of next neuron

Synapses In a relay race, the first runner sprints down the track with a baton in his or her hand. As the runner rounds the track, he or she hands the baton off to the next runner. The two runners never physically touch each other. The transfer of the baton signals the second runner to continue the race.

As shown in **Figure 3,** your nervous system works in a similar way. Like the runners in a relay race, neurons don't touch each other. How does an impulse move from one neuron to another? To move from one neuron to the next, an impulse crosses a small space called a **synapse** (SIH naps). In **Figure 4,** note that when an impulse reaches the end of an axon, the axon releases a chemical. This chemical flows across the synapse and stimulates the impulse in the dendrite of the next neuron. An impulse moves from neuron to neuron just like a baton moves from runner to runner in a relay race. The baton represents the chemical at the synapse.

The Central Nervous System

Figure 5 shows how organs of the nervous system are grouped into two major divisions—the central nervous system (CNS) and the peripheral (puh RIH fuh rul) nervous system (PNS). The **central nervous system** is made up of the brain and spinal cord. The **peripheral nervous system** is made up of all the nerves outside the CNS. These include the nerves in your head, called cranial nerves, and spinal nerves, which are nerves that come from your spinal cord. The peripheral nervous system connects the brain and spinal cord to other body parts. Your neurons are adapted in such a way that impulses move in only one direction. Sensory neurons send impulses to the brain or spinal cord.

Figure 5
The brain and spinal cord (yellow) form the central nervous system (CNS). All other nerves (green) are part of the peripheral nervous system (PNS).

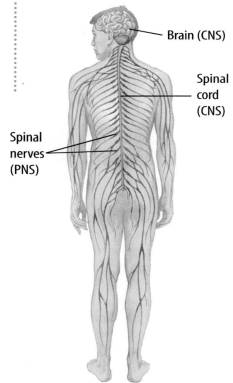

Brain (CNS)

Spinal cord (CNS)

Spinal nerves (PNS)

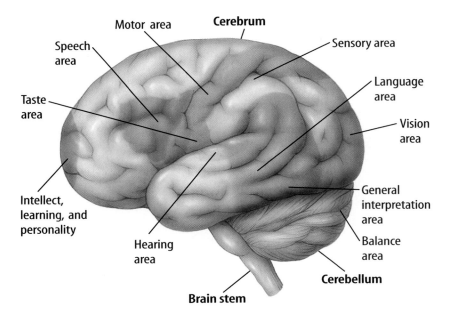

Motor area
Cerebrum
Speech area
Sensory area
Taste area
Language area
Vision area
Intellect, learning, and personality
General interpretation area
Balance area
Hearing area
Cerebellum
Brain stem

Figure 6
Different areas of the brain control specific body activities.

The Brain The brain coordinates all of your body activities. If someone tickles your feet, why does your whole body seem to react? The brain is made up of approximately 100 billion neurons, which is nearly ten percent of all the neurons in the human body. Surrounding and protecting the brain are a bony skull, three membranes, and a layer of fluid. As shown in **Figure 6,** the brain is divided into three major parts—the cerebrum (suh REE brum), the cerebellum (ser uh BEL um), and the brain stem.

Cerebrum Thinking takes place in the cerebrum. The **cerebrum** is the largest part of the brain. This is where impulses from the senses are interpreted, memory is stored, and movements are controlled. The outer layer of the cerebrum, called the cortex, is marked by many ridges and grooves. These structures increase the surface area of the cortex, allowing more complex thoughts to be processed. **Figure 6** shows some of the motor and sensory tasks that the cortex controls.

 **Reading Check** *What major activity takes place within the cerebrum?*

Cerebellum Stimuli from the eyes and ears and from muscles and tendons, which are the tissues that connect muscles to bones, are interpreted in the **cerebellum.** With this information, the cerebellum is able to coordinate voluntary muscle movements, maintain muscle tone, and help maintain balance. A complex activity, such as riding a bike, requires a lot of coordination and control of your muscles. The cerebellum coordinates your muscle movements so that you maintain your balance.

Brain Stem At the base of the brain is the **brain stem.** It extends from the cerebrum and connects the brain to the spinal cord. The brain stem is made up of the midbrain, the pons, and the medulla (muh DUH luh). The midbrain and pons act as pathways connecting various parts of the brain with each other. The medulla controls involuntary actions such as heartbeat, breathing, and blood pressure. The medulla also is involved in such actions as coughing, sneezing, swallowing, and vomiting.

Chemistry
INTEGRATION

Acetylcholine (uh see tul KOH leen) is a chemical produced by neurons, which carries an impulse across a synapse to the next neuron. After the impulse is started, the acetylcholine breaks down rapidly. In your Science Journal, hypothesize why the breakdown of acetylcholine is important.

The Spinal Cord Your spinal cord, illustrated in **Figure 7,** is an extension of the brain stem. It is made up of bundles of neurons that carry impulses from all parts of the body to the brain and from the brain to all parts of your body. The adult spinal cord is about the width of an adult thumb and is about 43 cm long.

The Peripheral Nervous System

Your brain and spinal cord are connected to the rest of your body by the peripheral nervous system. The PNS is made up of 12 pairs of nerves from your brain called cranial nerves, and 31 pairs from your spinal cord called spinal nerves. Spinal nerves are made up of bundles of sensory and motor neurons bound together by connective tissue. For this reason, a single spinal nerve can have impulses going to and from the brain at the same time. Some nerves contain only sensory neurons, and some contain only motor neurons, but most nerves contain both types of neurons.

Somatic and Autonomic Systems The peripheral nervous system has two major divisions. The somatic system controls voluntary actions. It is made up of the cranial and spinal nerves that go from the central nervous system to your skeletal muscles. The autonomic system controls involuntary actions—those not under conscious control—such as your heart rate, breathing, digestion, and glandular functions. These two divisions, along with the central nervous system, make up your body's nervous system.

Research Visit the Glencoe Science Web site at **science.glencoe.com** for more information about the nervous system. Make a brochure outlining recent medical advances.

Figure 7
Ⓐ A column of vertebrae, or bones, protects the spinal cord.
Ⓑ The spinal cord is made up of bundles of neurons that carry impulses to and from all parts of the body, similar to a telephone cable.

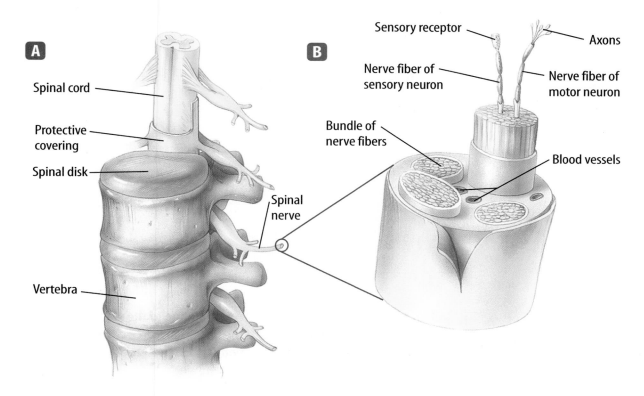

Safety and the Nervous System

Every mental process and physical action of the body is associated with the structures of the central and peripheral nervous systems. Therefore, any injury to the brain or the spinal cord can be serious. A severe blow to the head can bruise the brain and cause temporary or permanent loss of mental and physical abilities. For example, the back of the brain controls vision. An injury in this region could result in the loss of vision.

Although the spinal cord is surrounded by the bones in your spine called vertebrae, spinal cord injuries do occur. They can be just as dangerous as a brain injury. Injury to the spine can bring about damage to nerve pathways and result in paralysis (puh RAH luh suhs), which is the loss of muscle movement. As shown in **Figure 8,** a neck injury that damages certain nerves could prevent a person from breathing. Major causes of head and spinal injuries include automobile, motorcycle, and bicycle accidents, as well as sports injuries. Just like wearing seat belts in automobiles, it is important to wear the appropriate safety gear while playing sports and riding on bicycles and skateboards.

Figure 8
Head and spinal cord damage can result in paralysis depending on where the injury occurs.

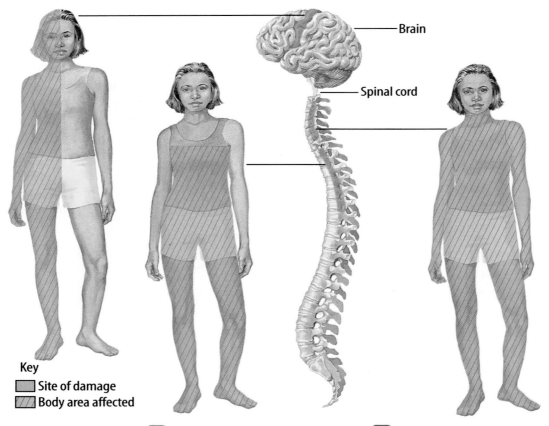

Brain

Spinal cord

Key
■ Site of damage
▨ Body area affected

A Damage to one side of the brain can result in the paralysis of the opposite side of the body.

B Damage to the middle or lower spinal cord can result in the legs and possibly part of the body being paralyzed.

C Damage to the spinal cord in the lower neck area can cause the body to be paralyzed from the neck down.

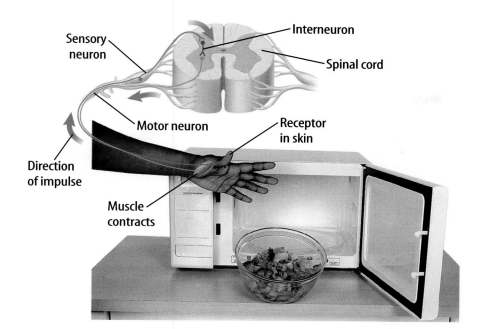

Sensory neuron

Interneuron

Spinal cord

Motor neuron

Receptor in skin

Direction of impulse

Muscle contracts

Figure 9
Your response in a reflex is controlled in your spinal cord, not your brain.

Reflexes You experience a reflex if you accidentally touch something sharp, something extremely hot or cold, or when you cough or vomit. A **reflex** is an involuntary, automatic response to a stimulus. You can't control reflexes because they occur before you know what has happened. A reflex involves a simple nerve pathway called a reflex arc, as illustrated in **Figure 9.**

While walking on a sandy beach, a pain suddenly shoots through your foot as you step on the sharp edge of a broken shell. Sensory receptors in your foot respond to this sharp object, and an impulse is sent to the spinal cord. As you just learned, the impulse passes to an interneuron in the spinal cord that immediately relays the impulse to motor neurons. Motor neurons transmit the impulse to muscles in your leg. Instantly, without thinking, you lift up your leg in response to the sharp-edged shell. This is a withdrawal reflex.

A reflex allows the body to respond without having to think about what action to take. Reflex responses are controlled in your spinal cord, not in your brain. Your brain acts after the reflex to help you figure out what to do to make the pain stop.

✔ Reading Check *Why are reflexes important?*

Do you remember reading at the beginning of this chapter about being frightened after a lamp was broken? What would have happened if your breathing and heart rate didn't calm down within a few minutes? Your body systems can't be kept in a state of continual excitement. The organs of your nervous system control and coordinate body responses. This helps maintain homeostasis within your body.

SCIENCE *Online*

Research Visit the Glencoe Science Web site at **science.glencoe.com** for more information about reflexes and paralysis. Make a small poster illustrating what you learn.

How Drugs Affect the Nervous System

Many drugs, such as alcohol and caffeine, directly affect your nervous system. When swallowed, alcohol directly passes through the walls of the stomach and small intestine into the circulatory system. After it is inside the circulatory system, it can travel throughout your body. Upon reaching neurons, alcohol moves through their cell membranes and disrupts their normal cell functions. As a result, this drug slows the activities of the central nervous system and is classified as a depressant. Muscle control, judgment, reasoning, memory, and concentration also are impaired. Heavy alcohol use destroys brain and liver cells.

Figure 10
Caffeine, a substance found in colas, coffee, chocolate, and some teas, can cause excitability and sleeplessness.

A stimulant is a drug that speeds up the activity of the central nervous system. Caffeine is a stimulant found in coffee, tea, cocoa, and many soft drinks, as shown in **Figure 10.** Too much caffeine can increase heart rate and aggravates restlessness, tremors, and insomnia in some people. It also can stimulate the kidneys to produce more urine.

Think again about a scare from a loud noise. The organs of your nervous system control and coordinate responses to maintain homeostasis within your body. This task might be more difficult when your body must cope with the effects of drugs.

Section Assessment

1. Draw and label the parts of a neuron.
2. Compare the central and peripheral nervous systems.
3. During a cold, winter evening, you have several cups of hot cocoa. Explain why you have trouble falling asleep that night.
4. Explain the advantage of having reflexes controlled by the spinal cord.
5. **Think Critically** Explain why many medications caution the consumer not to operate heavy machinery.

Skill Builder Activities

6. **Concept Mapping** Prepare an events-chain concept map of the different kinds of neurons that pass an impulse from a stimulus to a response. **For more help, refer to the** Science Skill Handbook.

7. **Using a Word Processor** Create a flowchart showing the reflex pathway of a nerve impulse when you step on a sharp object. Label the body parts involved in each step. **For more help, refer to the** Technology Skill Handbook.

Activity

Improving Reaction Time

Your reflexes allow you to react quickly without thinking. Sometimes you can improve how quickly you react. Complete this activity to see if you can decrease your reaction time.

What You'll Investigate
How can reaction time be improved?

Materials
metric ruler

Goals
■ **Observe** reflexes.
■ **Identify** stimuli and responses.

Procedure

1. Make a data table in your Science Journal to record where the ruler is caught during this activity. Possible column heads are Trial, Right Hand, and Left Hand.

2. Have a partner hold the ruler as shown.

3. Hold the thumb and index finger of your right hand apart at the bottom of the ruler. Do not touch the ruler.

4. Your partner must let go of the ruler without warning you.

5. Catch the ruler between your thumb and finger by quickly bringing them together.

6. Repeat this activity several times and record in a data table where the ruler was caught.

Communicating Your Data

Compare your conclusions with those of other students in your class. **For more help, refer to the** Science Skill Handbook.

7. Repeat this activity with your left hand.

Conclude and Apply

1. **Identify** the stimulus, response, and variable in this activity.

2. Use the table on the right to determine your reaction time.

3. What was the average reaction time for your right hand? For your left hand?

4. **Compare** the response of your writing hand and your other hand for this activity.

5. Draw a conclusion about how practice relates to stimulus-response time.

Reaction Time	
Where Caught (cm)	**Reaction Time (s)**
5	0.10
10	0.14
15	0.17
20	0.20
25	0.23
30	0.25

② The Senses

As You Read

What You'll Learn

- **List** the sensory receptors in each sense organ.
- **Explain** what type of stimulus each sense organ responds to and how.
- **Explain** why healthy senses are needed.

Vocabulary
retina
cochlea
olfactory cell
taste bud

Why It's Important
Your senses make you aware of your environment, enable you to enjoy your world, and help keep you safe.

The Body's Alert System

"Danger . . . danger . . . code red alert! An unidentified vessel has entered the spaceship's energy force field. All crew members are to be on alert!" Like spaceships in science fiction movies, your body has an alert system, too—your sense organs. You might see a bird, hear a dog bark, or smell popcorn. You can enjoy the taste of salt on a pretzel, the touch of a fuzzy peach, or feel heat from a warm, cozy fire. Light rays, sound waves, heat, chemicals, or pressure that comes into your personal territory will stimulate your sense organs. Sense organs are adapted for intercepting these different stimuli. They are then converted into impulses by the nervous system.

Vision

The eye, shown in **Figure 11,** is a sense organ. Think about the different kinds of objects you might look at every day. It's amazing that at one glance you might see the words on this page, the color illustrations, and your classmate sitting next to you. Your eyes have unique adaptations that usually enable you to see shapes of objects, shadows, and color.

Figure 11
Light moves through the cornea and the lens—before striking the retina.

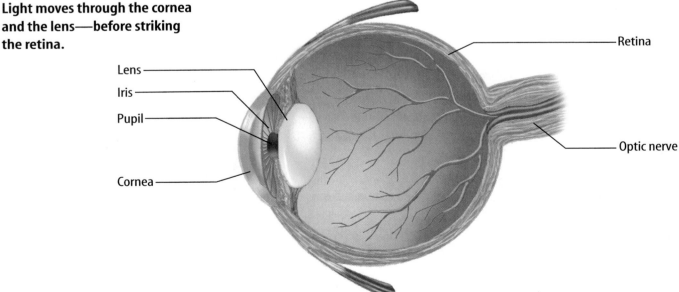

Lens

Iris

Pupil

Cornea

Retina

Optic nerve

How do you see? Light travels in a straight line unless something causes it to refract or change direction. Your eyes are equipped with structures that refract light. Two of these structures are the cornea and the lens. As light enters the eye, it passes through the cornea—the transparent section at the front of the eye—and is refracted. Then light passes through a lens and is refracted again. The lens directs the light onto the retina (RET nuh). The **retina** is a tissue at the back of the eye that is sensitive to light energy. Two types of cells called rods and cones are found in the retina. Cones respond to bright light and color. Rods respond to dim light. They are used to help you detect shape and movement. Light energy stimulates impulses in these cells.

The impulses pass to the optic nerve. This nerve carries the impulses to the vision area of the cortex, located on your brain's cerebrum. The image transmitted from the retina to the brain is upside down and reversed. The brain interprets the image correctly, and you see what you are looking at. The brain also interprets the images received by both eyes. It blends them into one image that gives you a sense of distance. This allows you to tell how close or how far away something is.

 **Reading Check** *What difficulties would a person who had vision only in one eye encounter?*

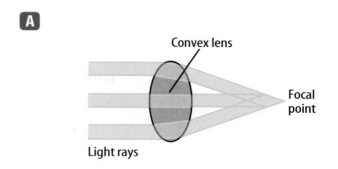

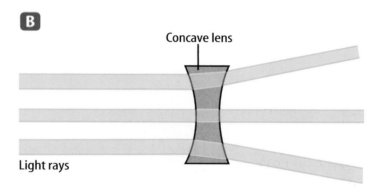

Figure 12
A Light passing through a convex lens is refracted toward the center and passes through a focal point. **B** Light that passes through a concave lens is refracted outward.

Lenses

 **Physics INTEGRATION**

Light is refracted when it passes through a lens. The way it refracts depends on the type of lens it passes through. A lens that is thicker in the middle and thinner on the edges is called a convex lens. As shown in **Figure 12A,** the lens in your eye refracts light so that it passes through a point, called a focal point. Convex lenses can be used to magnify objects. The light passes through a convex lens and enters the eye in such a way that your brain interprets the image as enlarged.

A lens that is thicker at its edges than in its middle is called a concave lens. Follow the light rays in **Figure 12B** as they pass through a concave lens. You'll see that this kind of lens causes the parallel light to spread out.

Figure 13
Glasses and contact lenses sharpen your vision.

A A nearsighted person cannot see distant objects because the image is focused in front of the retina.

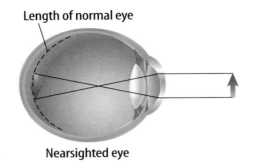

Length of normal eye

Nearsighted eye

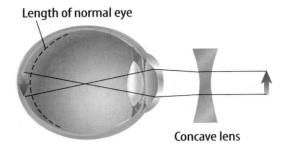

Length of normal eye

Concave lens

A concave lens corrects nearsightedness.

B A farsighted person cannot see close objects because the image is focused behind the retina.

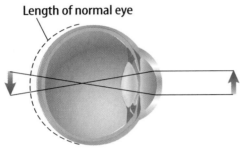

Length of normal eye

Farsighted eye

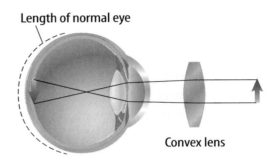

Length of normal eye

Convex lens

A convex lens corrects farsightedness.

Astronomy INTEGRATION

Refracting telescopes have two convex lenses for viewing objects in space. The larger lens collects light and forms an inverted, or upside-down, image of the object. The second lens magnifies the inverted image. In your Science Journal, hypothesize why telescopes used to view things on Earth have three lenses, not two.

Correcting Vision Problems Do you wear contact lenses or eyeglasses to correct your vision? Are you nearsighted or farsighted? In an eye with normal vision, light rays are focused onto the retina by the coordinated actions of the eye muscles, the cornea, and the lens. The image formed on the retina is interpreted by the brain as being sharp and clear. However, if the eyeball is too long from front to back, as illustrated in **Figure 13A,** light from objects is focused in front of the retina. This happens because the shape of the eyeball and lens cannot be changed enough by the eye muscles to focus a sharp image onto the retina. The image that reaches the retina is blurred. This condition is called nearsightedness—near objects are seen more clearly than distant objects. To correct nearsightedness, concave lenses are used to help focus images sharply on the retina.

Similarly, vision correction is needed when the eyeball is too short from front to back. In this case, light from objects is focused behind the retina despite the coordinated actions of the eye muscles, cornea, and lens. This condition is called farsightedness, as illustrated in **Figure 13B,** because distant objects are clearer than near objects. Convex lenses correct farsightedness.

Hearing

Whether it's the roar of a rocket launch, the cheers at a football game, or the distant song of a robin in a tree, sound waves are necessary for hearing sound. Sound energy is to hearing as light energy is to vision. When an object vibrates, sound waves are produced. These waves can travel through solids, liquids, and gases as illustrated in **Figure 14.** When the waves reach your ear, they usually stimulate nerve cells deep within your ear. Impulses are sent to the brain. When the sound impulse reaches the hearing area of the cortex, it responds and you hear a sound.

The Outer Ear and Middle Ear **Figure 15** shows that your ear is divided into three sections—the outer ear, middle ear, and inner ear. Your outer ear intercepts sound waves and funnels them down the ear canal to the middle ear. The sound waves cause the eardrum to vibrate much like the membrane on a musical drum vibrates when you tap it. These vibrations then move through three tiny bones called the hammer, anvil, and stirrup. The stirrup bone rests against a second membrane on an opening to the inner ear.

Figure 14
Objects produce sound waves that are heard by your ears.

Figure 15
Your ear responds to sound waves and to changes in the position of your head.

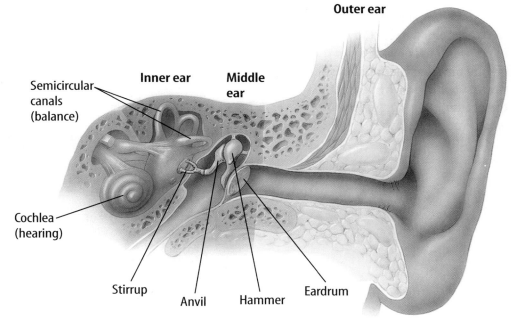

Outer ear

Inner ear

Middle ear

Semicircular canals (balance)

Cochlea (hearing)

Stirrup

Anvil

Hammer

Eardrum

Observing Balance Control

Procedure
1. Place **two narrow strips of paper** on the wall to form two parallel vertical lines 20–25 cm apart. Have a person stand between them for 3 min, without leaning on the wall.
2. Observe how well balance is maintained.
3. Have the person close his or her eyes, then stand within the lines for 3 min.

Analysis
1. When was balance more difficult to maintain? Why?
2. What other factors might cause a person to lose his or her sense of balance?

The Inner Ear The **cochlea** (KOH klee uh) is a fluid-filled structure shaped like a snail's shell. When the stirrup vibrates, fluids in the cochlea begin to vibrate. These vibrations bend hair cells in the cochlea, which causes electrical impulses to be sent to the brain by a nerve. High-pitched sounds make the endings move differently than lower sounds do. Depending on how the nerve endings are stimulated, you hear a different type of sound.

Balance Structures in your inner ear also control your body's balance. Structures called the cristae ampullaris (KRIHS tee • am pyew LEER ihs) and the maculae (MA kyah lee), illustrated in **Figure 16,** sense different types of body movement.

Both structures contain tiny hair cells. As your body moves, gel-like fluid surrounding the hair cells moves and stimulates the nerve cells at the base of the hair cells. This produces nerve impulses that are sent to the brain, which interprets the body movements. The brain, in turn, sends impulses to skeletal muscles, resulting in body movements that maintain balance.

The cristae ampullaris react to rotating body movements. Fluid in the semicircular canals swirls when the body rotates. This causes the gel-like fluid around the hair cells to move and a stimulus is sent to the brain. In a similar way, when the head tips, the gel-like fluid surrounding the hair cells in the maculae is pulled down by gravity. The hair cells are then stimulated and the brain interprets that the head has tilted.

Figure 16
Two structures in your inner ear are responsible for maintaining your sense of balance. **A** The cristae ampullaris react to rotating movements of your body. **B** The maculae check the position of your head with respect to the ground. *Why does spinning around make you dizzy?*

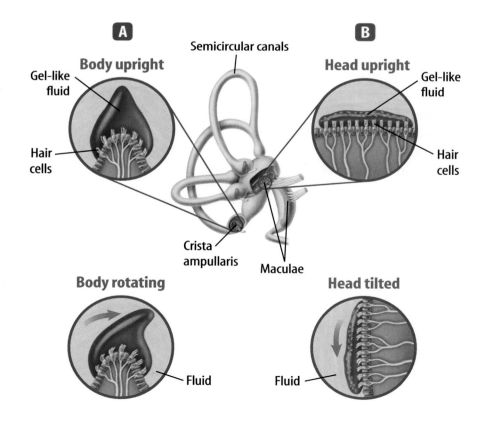

Smell

Some sharks can sense as few as ten drops of tuna liquid in an average-sized swimming pool. Even though your ability to detect odors is not as good as a shark's, your sense of smell is still important. Smell can determine which foods you eat. Strong memories or feelings also can be responses to something you smell.

You smell food because it gives off molecules into the air. These molecules stimulate sensitive nerve cells, called **olfactory** (ohl FAK tree) **cells,** in your nasal passages. Olfactory cells are kept moist by mucus. When molecules in the air dissolve in this moisture, the cells become stimulated. If enough molecules are present, an impulse starts in these cells, then travels to the brain where the stimulus is interpreted. If the stimulus is recognized from a previous experience, you may identify the odor. If you don't recognize a particular odor, it is remembered and may be identified the next time you encounter it.

SCIENCE *Online*

Research Visit the Glencoe Science Web site at **science.glencoe.com** for more information about the sense of smell in humans compared with other mammals. Make a chart in your Science Journal summarizing your research.

Math Skills Activity

Calculating Distance Using the Speed of Sound

Example Problem

You see the flash of fireworks and then four seconds later, you hear the boom because light waves travel faster than sound waves. Light travels so fast that you see it almost instantaneously. Sound, on the other hand, travels at 340 m/s. How far away are you from the source of the fireworks?

Solution

1 *This is what you know:*   time: $t = 4$ s
                                        speed of sound: $v = 340$ m/s

2 *This is what you need to find:*   distance: d

3 *This is the equation you need to use:* $d = vt$

4 *Substitute the known values* $d = (340 \text{ m/s})(4 \text{ s})$
                                          $d = 1360$ m

Check your answer by dividing your answer by time. Do you calculate the same speed that was given?

Practice Problem

A hiker standing at one end of a lake hears his echo 2.5 s after he shouts. It was reflected by a cliff at the end of the lake. How long is the lake?

For more help, refer to the Math Skill Handbook.

Mini LAB

Comparing Sense of Smell

Procedure

1. To test your classmates' abilities to recognize different odors, blindfold them one at a time, then pass near their noses small **samples of different foods, colognes, or household products. WARNING:** *Do not eat or drink anything in the lab. Do not use any products that give off noxious fumes.*
2. Ask each student to identify the different samples.
3. Record each student's response in a data table according to his or her gender.

Analysis

1. Compare the numbers of correctly identified odors for males and females.
2. What can you conclude about the differences between males and females in their abilities to recognize odors?

Taste

Sometimes you taste a new food with the tip of your tongue and find that it tastes sweet. Then when you chew it, you are surprised to find that it tastes bitter. **Taste buds** on your tongue are the major sensory receptors for taste. About 10,000 taste buds are found all over your tongue, enabling you to tell one taste from another.

Tasting Food Taste buds, shown in **Figure 17,** respond to chemical stimuli. When you think of hot french fries, your mouth begins to water. This response is helpful because in order to taste something, it has to be dissolved in water. Saliva begins this process. This solution of saliva and food washes over the taste buds, and impulses are sent to your brain. The brain interprets the impulses, and you identify the tastes. Most taste buds respond to several taste sensations. However, certain areas of the tongue are more receptive to one taste than another. The five taste sensations are sweet, salty, sour, bitter, and the taste of MSG (monosodium glutamate).

✔ **Reading Check** *What needs to happen to food before you are able to taste it?*

Smell and Taste Smell and taste are related. The sense of smell is needed to identify some foods such as chocolate. When saliva in your mouth mixes with the chocolate, odors travel up the nasal passage in the back of your throat. The olfactory cells are stimulated, and the taste and smell of chocolate are sensed. So when you have a stuffy nose and some foods seem tasteless, it may be because the food's molecules are blocked from contacting the olfactory cells in your nasal passages.

Figure 17
Taste buds are made up of a group of sensory cells with tiny taste hairs projecting from them. When food is taken into the mouth, it is dissolved in saliva. This mixture then stimulates receptor sites on the taste hairs, and an impulse is sent to the brain.

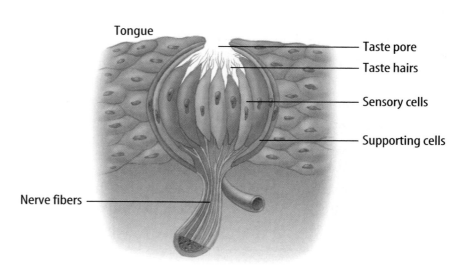

Tongue — Taste pore — Taste hairs — Sensory cells — Supporting cells — Nerve fibers

Other Sensory Receptors in the Body

As you are reading at school, you suddenly experience a bad pain in your lower right abdomen. The pain is not going away and you yell for help. Several hours later, you are resting in a hospital bed. The doctor has removed the source of your problem—your appendix. If not removed, a burst appendix can spread poison throughout your body.

Your internal organs have several kinds of sensory receptors. These receptors respond to touch, pressure, pain, and temperature. They pick up changes in touch, pressure, and temperature and transmit impulses to the brain or spinal cord. In turn, your body responds to this new information.

Sensory receptors also are located throughout your skin. As shown in **Figure 18,** your fingertips have many different types of receptors for touch. As a result, you can tell whether an object is rough or smooth, hot or cold, and hard or soft. Your lips are sensitive to heat and prevent you from drinking something so hot that it would burn you. Pressure-sensitive skin cells warn you of danger and enable you to move to avoid injury.

The body responds to protect itself from harm. All of your body's senses work together to maintain homeostasis. Your senses help you enjoy or avoid things around you. You constantly react to your environment because of information received by your senses.

Figure 18
Many of the sensations picked up by receptors in the skin are stimulated by mechanical energy. Pressure, motion, and touch are examples.

Section 2 Assessment

1. What type of stimulus do your ears respond to?

2. What are the sensory receptors for the eyes and nose?

3. Why is it important to have sensory receptors for pain and pressure in your internal organs?

4. What is the role of saliva in tasting?

5. **Think Critically** Unlike many other organs, the brain is insensitive to pain. What is the advantage of this?

Skill Builder Activities

6. **Making and Using Tables** Organize the information on senses in a table that names the sense organs and which stimuli they respond to. **For more help, refer to the** Science Skill Handbook.

7. **Communicating** Write a paragraph in your Science Journal that describes what each of the following objects would feel like: ice cube, snake, silk blouse, sandpaper, jelly, and smooth rock. **For more help, refer to the** Science Skill Handbook.

Activity

Skin Sensitivity

Your body responds to touch, pressure, temperature, and other stimuli. Not all parts of your body are equally sensitive to stimuli. Some areas are more sensitive than others are. For example, your lips are sensitive to heat. This protects you from burning your mouth and tongue. Now think about touch. How sensitive is the skin on various parts of your body to touch? Which areas can distinguish the smallest amount of distance between stimuli?

Recognize the Problem

What areas of the body are most sensitive to touch?

Form a Hypothesis

Based on your experiences, state a hypothesis about which of the following five areas of the body—fingertip, forearm, back of the neck, palm, and back of the hand—you believe to be most sensitive. Rank the areas from 5 (the most sensitive) to 1 (the least sensitive).

Goals
- **Observe** the sensitivity to touch on specific areas of the body.
- **Design** an experiment that tests the effects of a variable, such as how close the contact points are, to determine which body areas can distinguish which stimuli are closest to one another.

Possible Materials
3 × 5-inch index card
toothpicks
tape
*glue
metric ruler
*Alternate materials

Safety Precautions

Do not apply heavy pressure when touching the toothpicks to the skin of your classmates.

Test Your Hypothesis

Plan

1. As a group, agree upon and write the hypothesis statement.

2. As a group, list the steps you need to test your hypothesis. Describe exactly what you will do at each step. Consider the following as you list the steps. How will you know that sight is not a factor? How will you use the card shown on the right to determine sensitivity to touch? How will you determine that one or both points are sensed?

3. **Design** a data table in your Science Journal to record your observations.

4. Reread your entire experiment to make sure that all steps are in the correct order.

5. **Identify** constants, variables, and controls of the experiment.

Do

1. Make sure your teacher approves your plan before you start.

2. Carry out the experiment as planned.

3. While the experiment is going on, write down any observations that you make and complete the data table in your Science Journal.

Analyze Your Data

1. **Identify** which part of the body is least sensitive and which part is most sensitive.

2. **Identify** which part of the body tested can distinguish between the closest stimuli.

3. **Compare** your results with those of other groups.

4. Rank body parts tested from most to least sensitive. Did your results from this investigation support your hypothesis? Explain.

Draw Conclusions

1. Based on the results of your investigation, what can you infer about the distribution of touch receptors on the skin?

2. What other parts of your body would you predict to be less sensitive? Explain your predictions.

*C*ommunicating
Your Data

Write a report to share with your class about body parts of animals that are sensitive to touch. **For more help, refer to the** Science Skill Handbook.

Sula
by Toni Morrison

In the following passage from Sula, *a novel by Toni Morrison, the author describes Nel's response to the arrival of her old friend Sula.*

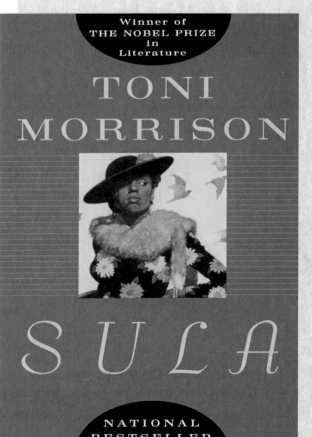

Winner of
THE NOBEL PRIZE
in
Literature

TONI
MORRISON

SULA

NATIONAL
BESTSELLER

Nel alone noticed the peculiar quality of the May that followed the leaving of the birds. It had a sheen, a glimmering as of green, rain-soaked Saturday nights (lit by the excitement of newly installed street lights); of lemon-yellow afternoons bright with iced drinks and splashes of daffodils. It showed in the damp faces of her children and the river-smoothness of their voices. Even her own body was not immune to the magic. She would sit on the floor to sew as she had done as a girl, fold her legs up under her or do a little dance that fitted some tune in her head. There were easy sun-washed days and purple dusks . . . .

Although it was she alone who saw this magic, she did not wonder at it. She knew it was all due to Sula's return to the Bottom.

Respond to the Reading

1. Describe, in your own words, how Nel feels about the return of her friend Sula.

2. What parts of the passage help you determine Nel's feelings?

Understanding Literature

Diction and Tone An author's choice of words, or diction, can help convey a certain tone in the writing. In the passage, Toni Morrison begins by describing a day in May. Her choice of words—like *sheen, glimmering, lemon-yellow afternoons,* and *splashes of daffodils*—conveys a happy or pleasant tone. These word choices help the reader understand that the character Nel is enjoying the month of May. Many other examples found in the passage show how the author's diction helps convey a happy or pleasant tone. Find two more examples in which diction conveys a pleasant tone.

Science Connection In this chapter, you learned how the body and its nervous system react to stimuli in the environment. In the passage you just read, Nel has a physical reaction to her environment. She is moved to "do a little dance" in response to the sights and sounds of May. This action is an example of a voluntary response to stimuli from outside the body. Movement of the body is a coordinated effort of the skeletal, muscular, and nervous system. Nel can dance because motor neurons conduct impulses from the brain to her muscles.

Linking Science and Writing

Choosing Words to Convey a Tone Write a paragraph describing the month of January that clearly shows a person's dislike for the month. Think about the month of January and the physical reactions people have to their surroundings during this time of the year. Convey your character's dislike through the description of his or her nervous system's response to stimuli from the January environment. For example, you might say that your character shivers in the harsh wind. The trick is to do this without directly saying that your character does not like January.

Career Connection

Anthropologist

Katherine Mary Dunham has transferred her knowledge and experience with control and coordination into two careers. She is a dance choreographer as well as an anthropologist, which is someone who studies the origins of the physical, social, and cultural development of human beings. She received a master's degree in science from the University of Chicago and a doctoral degree from Northwestern University. Her research in these two fields led her to the development of an African-based theory of movement. She has created a training center in St. Louis, Missouri where she teaches inner-city youths African culture and dance.

SCIENCE *Online* To learn more about careers in anthropology, visit the Glencoe Science Web site at **science.glencoe.com.**

Reviewing Main Ideas

Section 1 The Nervous System

1. Your body constantly is receiving a variety of stimuli from inside and outside the body. The nervous system responds to these stimuli to maintain homeostasis.

2. A neuron is the basic unit of structure and function of the nervous system.

3. A stimulus is detected by sensory neurons. Electrical impulses are carried to the interneurons and transmitted to the motor neurons. The result is the movement of a body part. *What are some body functions that are being checked and regulated constantly?*

4. A response that is made automatically is a reflex.

5. The central nervous system contains the brain and spinal cord. The peripheral nervous system is made up of cranial and spinal nerves.

6. Many drugs, such as alcohol and caffeine, have a direct effect on your nervous system. *What are some effects of caffeine found in foods such as chocolate?*

Section 2 The Senses

1. Your senses respond to stimuli. The eyes respond to light energy, and the ears respond to sound waves.

2. Olfactory cells of the nose and taste buds of the tongue are stimulated by chemicals.

3. Sensory receptors in your internal organs and skin respond to touch, pressure, pain, and temperature.

4. Your senses enable you to enjoy or avoid things around you. You are able to react to the changing conditions of your environment. *What senses are involved when you pick up and eat a freshly baked piece of pita bread filled with delicious ingredients?*

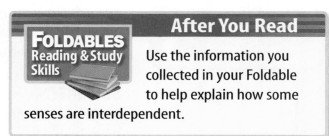

FOLDABLES
Reading & Study Skills

After You Read

Use the information you collected in your Foldable to help explain how some senses are interdependent.

Visualizing Main Ideas

Examine the following concept map of the nervous system and fill in the missing terms.

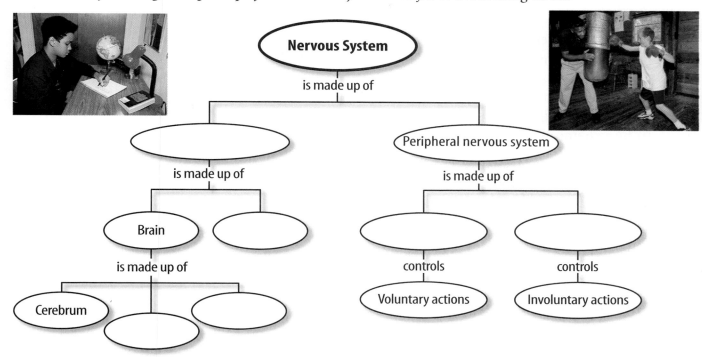

Vocabulary Review

Vocabulary Words

a. axon
b. brain stem
c. central nervous system
d. cerebellum
e. cerebrum
f. cochlea
g. dendrite
h. homeostasis
i. neuron
j. olfactory cell
k. peripheral nervous system
l. reflex
m. retina
n. synapse
o. taste bud

Study Tip

THE PRINCETON REVIEW

Use word webs. Write the main idea of the chapter on a piece of paper and circle it. Connect other related facts to it with lines and arrows.

Using Vocabulary

Explain the difference between the vocabulary words in each of the following sets.

1. axon, dendrite

2. central nervous system, peripheral nervous system

3. cerebellum, cerebrum

4. reflex, synapse

5. brain stem, neuron

6. olfactory cell, taste bud

7. dendrite, synapse

8. cerebrum, central nervous system

9. retina, cochlea

10. synapse, neuron

Checking Concepts

Choose the word or phrase that best answers the question.

1. How do impulses cross synapses between neurons?
   A) by osmosis
   B) through interneurons
   C) through a cell body
   D) by a chemical

2. What are the neuron structures that carry impulses to the cell body called?
   A) axons
   B) dendrites
   C) synapses
   D) nuclei

3. What are neurons called that detect stimuli in the skin and eyes?
   A) interneurons
   B) motor neurons
   C) sensory neurons
   D) synapses

4. Which of the following does the skin not sense?
   A) pain
   B) pressure
   C) temperature
   D) taste

5. What part of the brain controls voluntary muscles?
   A) cerebellum
   B) brain stem
   C) cerebrum
   D) pons

6. What part of the brain has an outer layer called the cortex?
   A) pons
   B) brain stem
   C) cerebrum
   D) spinal cord

7. What does the somatic system of the PNS control?
   A) skeletal muscles
   B) heart
   C) glands
   D) salivary glands

8. What part of the eye is light finally focused on?
   A) lens
   B) retina
   C) pupil
   D) cornea

9. What is the largest part of the brain?
   A) cerebellum
   B) brain stem
   C) cerebrum
   D) pons

10. Which of the following is in the inner ear?
    A) anvil
    B) hammer
    C) eardrum
    D) cochlea

Thinking Critically

11. Why is it helpful to have impulses move only in one direction in a neuron?

12. How are reflexes protective?

13. Describe how smell and taste are related.

14. How does the use of alcohol influence a person's ability to drive a car?

15. If a fly were to land on your face and another one on your back, which might you feel first? How could you test your choice?

Developing Skills

16. **Concept Mapping** Fill in this events-chain concept map that shows the correct sequence of the structures through which light passes in the eye.

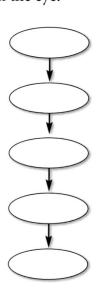

17. Classifying Group the types of neurons as to their location and direction of impulse.

18. Comparing and Contrasting Compare and contrast the structures and functions of the cerebrum, cerebellum, and brain stem. Include in your discussion the following functions: balance, involuntary muscle movements, muscle tone, memory, voluntary muscles, thinking, and senses.

19. Drawing Conclusions If an impulse traveled down one neuron but failed to move on to the next neuron, what might you conclude about the first neuron?

20. Interpreting Scientific Illustrations Using the following diagram, explain how an impulse crosses a synapse.

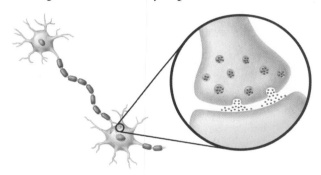

Performance Assessment

21. Illustrate In an emergency room, the doctor notices that a patient has uncoordinated body movements and has difficulty maintaining his balance. Draw and label which part of the brain may be injured.

TECHNOLOGY

Go to the Glencoe Science Web site at **science.glencoe.com** or use the **Glencoe Science CD-ROM** for additional chapter assessment.

 **Test Practice**

A police officer brought the following table into a school to educate students about the dangers of drinking and driving.

Approximate Blood Alcohol Percentage for Men								
Drinks	Body Weight in Kilograms							
	45.4	54.4	63.5	72.6	81.6	90.7	99.8	108.9
1	0.04	0.03	0.03	0.02	0.02	0.02	0.02	0.02
2	0.08	0.06	0.05	0.05	0.04	0.04	0.03	0.03
3	0.11	0.09	0.08	0.07	0.06	0.06	0.05	0.05
4	0.15	0.12	0.11	0.09	0.08	0.08	0.07	0.06
5	0.19	0.16	0.13	0.12	0.11	0.09	0.09	0.08

Subtract 0.01% for each 40 minutes of drinking. One drink is 40 mL of 80 proof liquor, 355 mL of beer, or 148 mL of table wine.

Study the table and answer the questions.

1. In some states, the legal blood alcohol percentage limit for driving while under the influence of alcohol is 0.08 percent. According to this information, how many drinks would it take for a 99-kg man to exceed this limit?
A) three   **C)** five
B) four   **D)** six

2. A 72-kg man has been tested for blood alcohol content. His blood alcohol percentage is 0.07. Based upon the information in the table, about how much has he had to drink?
F) 628 mL of 80-proof liquor
G) 1,064 mL of beer
H) 295 mL of table wine
J) four drinks

Regulation and Reproduction

The control room blinks with monitors and panels of dials and buttons. Not much is going to get past this complex monitoring system. Your body also is designed with a system that monitors and controls the actions of many of your body's functions. In this chapter, you'll learn about this system—the endocrine system. You'll also study the human reproductive system and the stages of growth.

What do you think?

Science Journal Look at the picture below with a classmate. Discuss what this might be. Here's a hint: *This object could be considered a small beginning.* Write your answer or best guess in your Science Journal.

Your body has systems that work together to control your body's activities. One of these systems sends chemical messages through your blood to certain tissues, which, in turn, respond. You may feel the results of this system's action, but you cannot see them. Do the activity below to see how a chemical signal can be sent.

Model a chemical message

1. Cut a 10-cm-tall Y shape from filter paper and place it on a plastic, ceramic, or glass plate.

2. Sprinkle baking soda on one arm of the Y and salt on the other arm.

3. Using a dropper, place five or six drops of vinegar halfway up the leg of the Y.

Observe

Describe in your Science Journal how the chemical moves along the paper and the reaction(s) it causes.

Before You Read

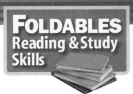

FOLDABLES
Reading & Study Skills

Making a Sequence Study Fold
Make the following Foldable to help you predict what might occur next in the sequence of life.

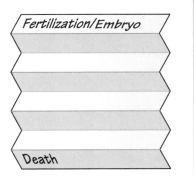

Fertilization/Embryo

Death

1. Place a sheet of paper in front of you so the short side is at the top. Fold the paper in half from top to bottom. Then fold it in half again top to bottom two more times. Unfold all the folds.

2. Using the fold lines as a guide, refold the paper into a fan. Unfold all the folds again.

3. Before you read the chapter list as many stages of life as you can on your foldable, beginning with *Fertilization/Embryo* and ending with *Death*. As you read the chapter add to your list.

SECTION 1

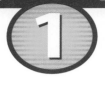

The Endocrine System

As You Read

What You'll Learn

- **Define** how hormones function.
- **Identify** different endocrine glands and the effects of the hormones they produce.
- **Describe** how a feedback system works in your body.

Vocabulary
hormone

Why It's Important
The endocrine system uses chemicals to control many systems in your body.

Functions of the Endocrine System

You go through the dark hallways of a haunted house. You can't see a thing. Your heart is pounding. Suddenly, a monster steps out in front of you. You scream and jump backwards. Your body is prepared to defend itself or get away. Preparing the body for fight or flight in times of emergency, as shown in **Figure 1,** is one of the functions of the body's control systems.

Chemical Messengers Your body is made up of systems that are controlled, or regulated, to work together. The nervous system and the endocrine (EN duh krun) system are the control systems of your body. The nervous system sends messages to and from the brain throughout the body. The endocrine system uses **hormones** (HOR mohnz)—chemicals that are made in tissues called glands found throughout your body. Hormones from endocrine glands are released directly into your bloodstream. They affect specific tissues called target tissues, usually located in the body far from the hormone-producing gland. The body doesn't react as quickly to messages from the endocrine system as it does to those of the nervous system.

Figure 1
Your endocrine system enables many parts of your body to respond with an immediate reaction to a fearful situation.

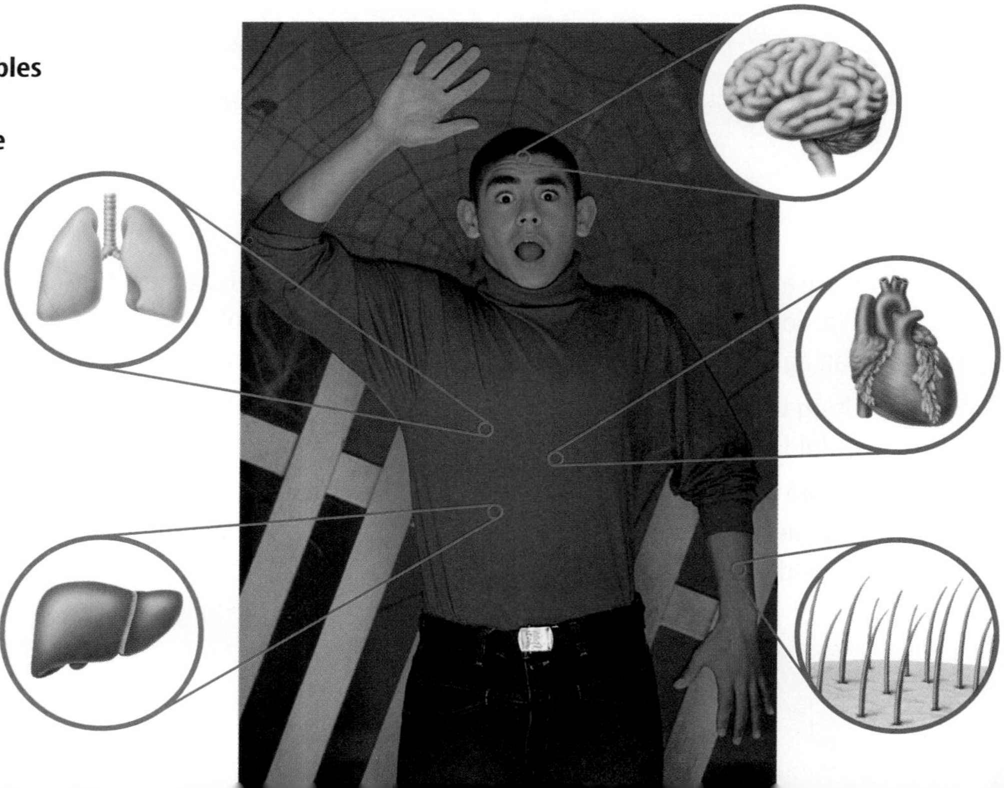

Endocrine Glands

Unlike some glands, such as your mouth's saliva glands, that release their products through small tubes called ducts, endocrine glands are ductless. Hormones from endocrine glands pour directly into the blood to reach target tissues. Hormones regulate certain cellular activities. **Figure 2** on the next page describes some of your major endocrine glands and how they function to regulate your body.

 Reading Check *What is the function of hormones?*

Earth Science
INTEGRATION

Without the element iodine, the thyroid gland cannot function properly. Iodine is found in seawater, soil, and rocks. How does your body take in iodine? Write your answer in your Science Journal.

Math Skills Activity

Calculating Blood Sugar Percentage

Example Problem

Calculate how much higher the blood sugar (glucose) level of a diabetic is before breakfast when compared to a nondiabetic before breakfast. Express this number as a percentage of the nondiabetic sugar level before breakfast.

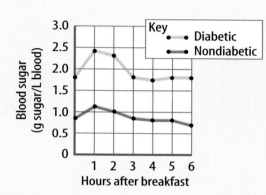

Solution

1 *This is what you know:*
blood sugar of a nondiabetic person at 0 h = 0.85 g sugar/L blood
blood sugar of a diabetic person at 0 h = 1.8 g sugar/L blood

2 *This is what you must do first:*
Find the difference between the two values. 1.8 g/L − 0.85 g/L = 0.95 g/L

3 *This is the equation you need to use:*

$$\frac{\text{difference between values}}{\text{nondiabetic value}} \times 100\% = \text{percent difference}$$

4 *Substitute in the known values:*

$$\frac{0.95}{0.85} \times 100\% = 111\%$$

At 0 h before breakfast, a diabetic's blood sugar is 111 percent higher than that of a nondiabetic.

Practice Problem

Express as a percentage how much higher the blood sugar value is for a diabetic person compared to a nondiabetic person 1 h, 3 h, and 6 h after breakfast.

For more help, refer to the Math Skill Handbook.

Figure 2

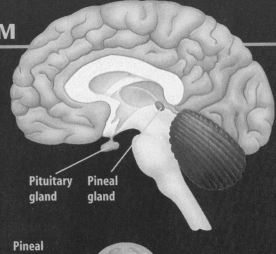

Your endocrine system is involved in regulating and coordinating many body functions, from growth and development to reproduction. This complex system consists of many diverse glands and organs, including the nine shown here. Endocrine glands produce chemical messenger molecules, called hormones, that circulate in the bloodstream. Hormones exert their influence only on the specific target cells to which they bind.

PINEAL GLAND Shaped like a tiny pine cone, the pineal gland lies deep in the brain. It produces melatonin, a hormone that may function as a sort of body clock by regulating wake/sleep patterns.

PITUITARY GLAND A pea-size structure attached to the hypothalamus of the brain, the pituitary gland produces hormones that affect a wide range of body activities, from growth to reproduction.

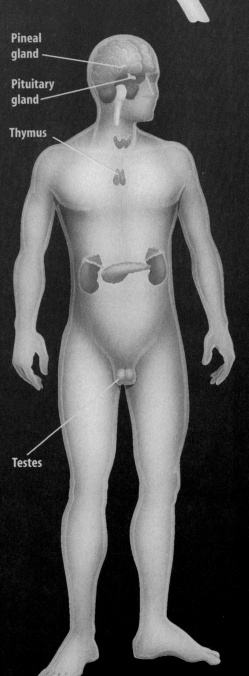

THYMUS The thymus is located in the upper chest, just behind the sternum. Hormones produced by this organ stimulate the production of certain infection-fighting cells.

TESTES These paired male reproductive organs primarily produce testosterone, a hormone that controls the development and maintenance of male sexual traits. Testosterone also plays an important role in the production of sperm.

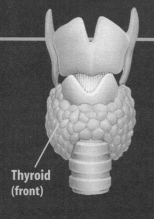

THYROID GLAND Located below the larynx, the bi-lobed thyroid gland is richly supplied with blood vessels. It produces hormones that regulate metabolic rate, control the uptake of calcium by bones, and promote normal nervous system development.

Thyroid (front)

PARATHYROID GLANDS Attached to the back surface of the thyroid are tiny para-thyroids, which help regulate calcium levels in the body. Calcium is important for bone growth and maintenance, as well as for muscle contraction and nerve impulse transmission.

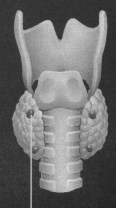

Parathyroid (back)

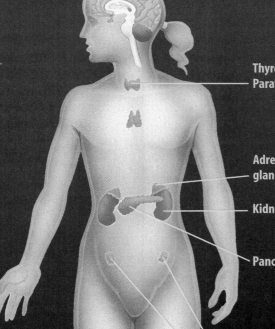

Thyroid and Parathyroid

Adrenal gland

Kidney

Pancreas

Ovaries

ADRENAL GLANDS On top of each of your kidneys is an adrenal gland. This complex endocrine gland produces a variety of hormones. Some play a critical role in helping your body adapt to physical and emotional stress. Others help stabilize blood sugar levels.

PANCREAS Scattered throughout the pancreas are millions of tiny clusters of endocrine tissue called the islets of Langerhans. Cells that make up the islets produce hormones that help control sugar levels in the bloodstream.

OVARIES Found deep in the pelvic cavity, ovaries produce female sex hormones known as estrogen and progesterone. These hormones regulate the female reproductive cycle and are responsible for producing and maintaining female sex characteristics.

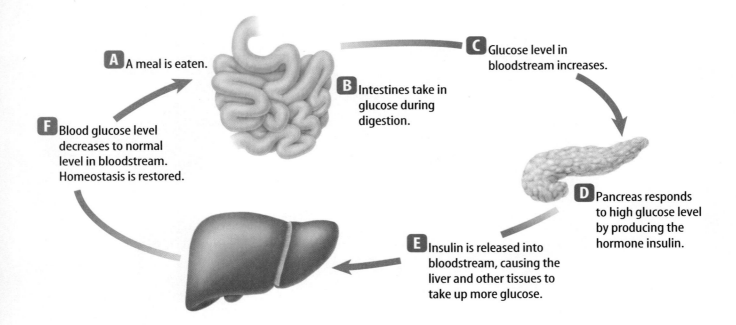

A A meal is eaten.

B Intestines take in glucose during digestion.

C Glucose level in bloodstream increases.

D Pancreas responds to high glucose level by producing the hormone insulin.

E Insulin is released into bloodstream, causing the liver and other tissues to take up more glucose.

F Blood glucose level decreases to normal level in bloodstream. Homeostasis is restored.

Figure 3
Many internal body conditions, such as hormone level, blood sugar level, and body temperature, are controlled by negative-feedback systems. Using a negative-feedback system, the pancreas controls the level of glucose in your bloodstream.

A Negative-Feedback System

To control the amount of hormones that are in your body, the endocrine system sends chemical messages back and forth within itself. This is called a negative-feedback system. It works much the way a thermostat works. When the temperature in a room drops below a set level, the thermostat signals the furnace to turn on. Once the furnace has raised the temperature in the room to the set level, the thermostat signals the furnace to shut off. It will continue to stay off until the thermostat signals that the temperature has dropped again. **Figure 3** shows how a negative-feedback system controls the level of glucose in your bloodstream.

Section ① Assessment

1. Compare and contrast the human body's two control systems.
2. What is the function of hormones?
3. Choose one endocrine gland and explain how it works.
4. What is a negative-feedback system?
5. **Think Critically** Glucose is required for cellular respiration, the process that releases energy within cells. How would lack of insulin affect this process?

Skill Builder Activities

6. **Predicting** Predict why the circulatory system is a good mechanism for delivering hormones throughout the body. **For more help, refer to the** Science Skill Handbook.

7. **Researching Information** Research recent treatments for growth disorders involving the pituitary gland. Write a brief paragraph of your results in your Science Journal. **For more help, refer to the** Science Skill Handbook.

The Reproductive System

Reproduction and the Endocrine System

Reproduction is the process that continues life on Earth. Most human body systems, such as the digestive system and the nervous system, are the same in males and females, but this is not true for the reproductive system. Males and females each have structures specialized for their roles in reproduction. Although structurally different, both the male and female reproductive systems are adapted to allow for a series of events that can lead to the birth of a baby.

Hormones are the key to how the human reproductive system functions, as shown in **Figure 4.** Sex hormones are necessary for the development of sexual characteristics, such as breast development in females and facial hair growth in males. Hormones from the pituitary gland also begin the production of eggs in females and sperm in males. Eggs and sperm transfer hereditary information from one generation to the next.

As You Read

What You'll Learn
- **Identify** the function of the reproductive system.
- **Compare and contrast** the major structures of the male and female reproductive systems.
- **Sequence** the stages of the menstrual cycle.

Vocabulary

testes	uterus
sperm	vagina
semen	menstrual cycle
ovary	menstruation
ovulation	

Why It's Important
The reproductive system helps ensure that life continues on Earth.

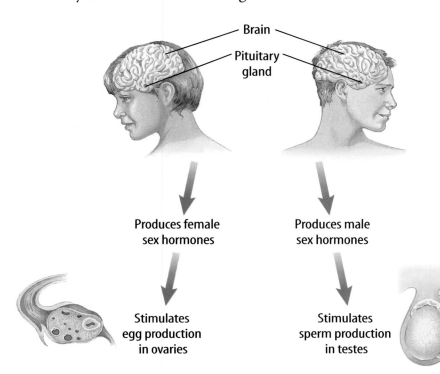

Brain

Pituitary gland

Produces female sex hormones

Stimulates egg production in ovaries

Produces male sex hormones

Stimulates sperm production in testes

Figure 4
The pituitary gland produces hormones that control the male and female reproductive systems.

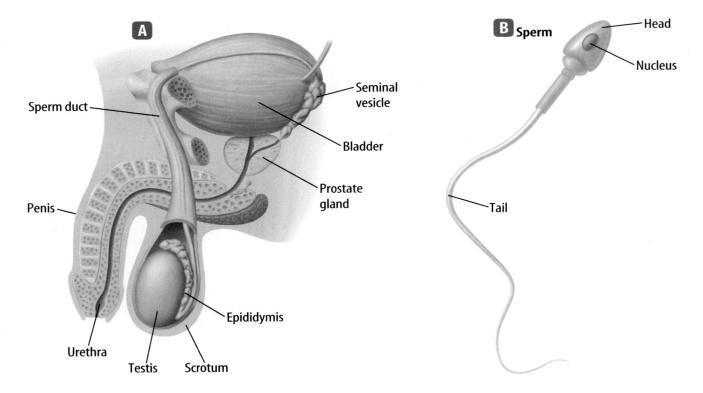

A

Sperm duct

Seminal vesicle

Bladder

Penis

Prostate gland

Epididymis

Urethra

Testis   Scrotum

B Sperm

Head

Nucleus

Tail

Figure 5
The structures of **A** the male reproductive system are shown with **B** a close-up of the sperm, which is produced in the testis. Sperm are produced throughout the life of a male.

The Male Reproductive System

The male reproductive system is made up of external and internal organs. The external organs of the male reproductive system are the penis and scrotum, shown in **Figure 5.** The scrotum contains two organs called testes (TES teez). As males mature sexually, the **testes** begin to produce testosterone, the male hormone, and **sperm,** which are male reproductive cells.

Sperm Each sperm cell has a head and tail. The head contains hereditary information, and the tail moves the sperm. Because the scrotum is located outside the body cavity, the testes, where sperm are produced, are kept at a lower temperature than the rest of the body. Sperm are produced in greater numbers at lower temperatures.

Many organs help in the production, transportation, and storage of sperm. After sperm are produced, they travel from the testes through sperm ducts that circle the bladder. Behind the bladder, a gland called the seminal vesicle provides sperm with a fluid. This fluid supplies the sperm with an energy source and helps them move. This mixture of sperm and fluid is called **semen** (SEE mun). Semen leaves the body through the urethra, which is the same tube that carries urine from the body. However, semen and urine never mix. A muscle at the back of the bladder contracts to prevent urine from entering the urethra as sperm leave the body.

The Female Reproductive System

Unlike male reproductive organs, most of the reproductive organs of the female are inside the body. The **ovaries**—the female sex organs—are located in the lower part of the body cavity. Each of the two ovaries is about the size and shape of an almond. **Figure 6** shows the different organs of the female reproductive system.

The Egg When a female is born, she already has all of the cells in her ovaries that eventually will develop into eggs—the female reproductive cells. At puberty, eggs start to develop in her ovaries because of specific sex hormones.

About once a month, an egg is released from an ovary in a hormone-controlled process called **ovulation** (ahv yuh LAY shun). The two ovaries release eggs on alternating months. One month, an egg is released from an ovary. The next month, the other ovary releases an egg and so on. After the egg is released, it enters the oviduct. Sometimes a sperm fertilizes the egg. If fertilization takes place, it usually happens in an oviduct. Short, hairlike structures called cilia help sweep the egg through the oviduct toward the uterus (YEWT uh rus).

✔ **Reading Check** *When are eggs released by the ovaries?*

The **uterus** is a hollow, pear-shaped, muscular organ with thick walls in which a fertilized egg develops. The lower end of the uterus, the cervix, narrows and is connected to the outside of the body by a muscular tube called the **vagina** (vuh JI nuh). The vagina also is called the birth canal because during birth, a baby travels through this tube from the uterus to the outside of the mother's body.

SCIENCE *Online*

Research Visit the Glencoe Science Web site at **science. glencoe. com** for information about ovarian cysts. Make a small pamphlet explaining what cysts are and how they can be treated.

Figure 6
The structures of the female reproductive system are shown from the A side of the body and from the B front. *Where in the female reproductive system do the eggs develop?*

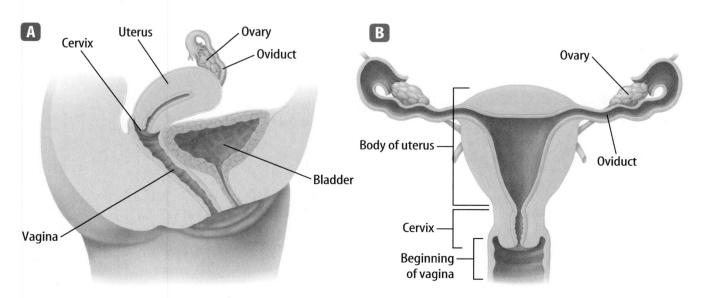

Graphing Hormone Levels

Procedure
Make a line graph of this table.

Hormone Changes	
Day	Level of Hormone
1	12
5	14
9	15
13	70
17	13
21	12
25	8

Analysis
1. On what day is the highest level of hormone present?
2. What event takes place around the time of the highest hormone level?

The Menstrual Cycle

How is the female body prepared for having a baby? The **menstrual cycle** is the monthly cycle of changes in the female reproductive system. Before and after an egg is released from an ovary, the uterus undergoes changes. The menstrual cycle of a human female averages 28 days. However, the cycle can vary in some individuals from 20 to 40 days. Changes include the maturing of an egg, the production of female sex hormones, and the preparation of the uterus to receive a fertilized egg.

✔ **Reading Check** *What is the menstrual cycle?*

Endocrine Control Hormones control the entire menstrual cycle. The pituitary gland responds to chemical messages from the hypothalamus by releasing several hormones. These hormones start the development of eggs in the ovary. They also start the production of other hormones in the ovary, including estrogen (ES truh jun) and progesterone (proh JES tuh rohn). The interaction of all these hormones results in the physical processes of the menstrual cycle.

Phase One As shown in **Figure 7,** the first day of phase 1 starts when menstrual flow begins. Menstrual flow consists of blood and tissue cells released from the thickened lining of the uterus. This flow usually continues for four to six days and is called **menstruation** (men STRAY shun).

Figure 7
The three phases of the menstrual cycle make up the monthly changes in the female reproductive system.

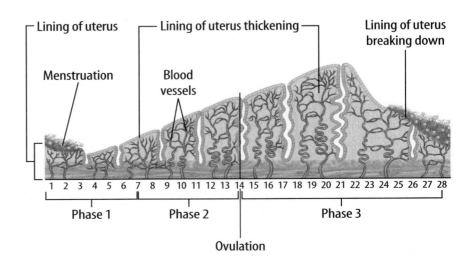

Phase Two Hormones cause the lining of the uterus to thicken in phase 2. Hormones also control the development of an egg in the ovary. Ovulation occurs about 14 days before menstruation begins. Once the egg is released, it must be fertilized within 24 h or it usually begins to break down. Because sperm can survive in a female's body for up to three days, fertilization can occur soon after ovulation.

Phase Three Hormones produced by the ovaries continue to cause an increase in the thickness of the uterine lining during phase 3. If a fertilized egg does arrive, the uterus is ready to support and nourish the developing embryo. If the egg is not fertilized, the lining of the uterus breaks down as the hormone levels decrease. Menstruation begins and the cycle repeats itself.

Menopause For most females the first menstrual period happens between ages nine years and 13 years and continues until 45 years of age to 60 years of age. Then, a gradual reduction of menstruation takes place as hormone production by the ovaries begins to shut down. Menopause occurs when both ovulation and menstrual periods end. It can take several years for the completion of menopause. As **Figure 8** indicates, menopause does not inhibit a woman's ability to enjoy an active life.

Figure 8
This older woman enjoys exercising with her granddaughter.

Section 2 Assessment

1. What is the major function of male and female reproductive systems in humans?
2. Explain the movement of sperm through the male reproductive system.
3. Compare and contrast the major organs and structures of the male and female reproductive systems.
4. Using diagrams and captions, sequence the stages of the menstrual cycle in a human female.
5. **Think Critically** Adolescent females often require additional amounts of iron in their diet. Explain.

Skill Builder Activities

6. **Concept Mapping** Make an events chain concept map to sequence the movement of an egg through the female reproductive system. **For more help, refer to the** Science Skill Handbook.
7. **Solving One-Step Equations** Usually, one egg is released each month during a female's reproductive years. If menstruation begins at 12 years of age and ends at 50 years of age, calculate the possible number of eggs her body can release during her reproductive years. **For more help, refer to the** Math Skill Handbook.

Activity

Interpreting Diagrams

Starting in adolescence, hormones cause the development of eggs in the ovary and changes in the uterus. These changes prepare the uterus to accept a fertilized egg that can attach itself in the wall of the uterus. What happens to an unfertilized egg?

What You'll Investigate
What changes occur to the uterus during a female's monthly menstrual cycle?

Materials
paper and pencil

Goals
- ■ **Observe** the stages of the menstrual cycle in the diagram.
- ■ **Relate** the process of ovulation to the cycle.

Menstruation Cycle		
Days	**Condition of Uterus**	**What Happens**
1–6		
7–12		
13–14		
15–18		

Conclude and Apply
1. How many days does the average menstrual cycle last?
2. On what days does the lining of the uterus build up?
3. **Infer** why this process is called a cycle.
4. **Calculate** how many days before menstruation ovulation usually occurs.

Procedure
1. The diagrams below show what is explained in this chapter on the menstrual cycle.
2. Use the information in this chapter and the diagrams below to complete a data table.
3. On approximately what day in a 28-day cycle is the egg released from the ovary?

*C*ommunicating
Your Data

Compare your data table with those of other students in your class. **For more help, refer to the** Science Skill Handbook.

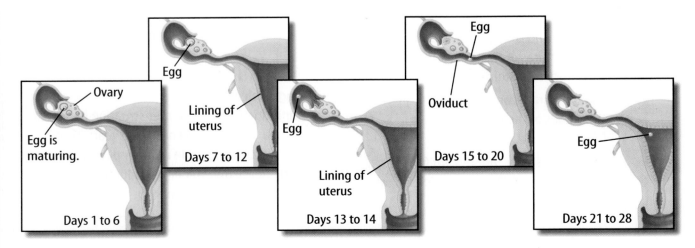

Ovary

Egg is maturing.

Days 1 to 6

Egg

Lining of uterus

Days 7 to 12

Egg

Lining of uterus

Days 13 to 14

Egg

Oviduct

Days 15 to 20

Egg

Days 21 to 28

Human Life Stages

The Function of the Reproductive System

Before the invention of powerful microscopes, some people imagined an egg or a sperm to be a tiny person that grew inside a female. In the latter part of the 1700s, experiments using amphibians showed that contact between an egg and sperm is necessary for the development of life. With the development of the cell theory in the 1800s, scientists recognized that a human develops from an egg that has been fertilized by a sperm. The uniting of a sperm and an egg is known as fertilization. Fertilization, as shown in **Figure 9,** usually takes place in the oviduct.

Fertilization

Although 200 millon to 300 million sperm can be deposited in the vagina, only several thousand reach an egg in the oviduct. As they enter the female, the sperm come into contact with chemical secretions in the vagina. It appears that this contact causes a change in the membrane of the sperm. The sperm then become capable of fertilizing the egg. The one sperm that makes successful contact with the egg releases an enzyme from the saclike structure on its head. Enzymes help speed up chemical reactions that have a direct effect on the protective membranes on the egg's surface. The structure of the egg's membrane is disrupted, and the sperm head can enter the egg.

Zygote Formation Once a sperm has entered the egg, changes in the electric charge of the egg's membrane prevent other sperm from entering the egg. At this point, the nucleus of the successful sperm joins with the nucleus of the egg. This joining of nuclei creates a fertilized cell called the zygote. It begins to undergo many cell divisions.

Figure 9
After the sperm releases enzymes that disrupt the egg's membrane, it penetrates the egg.

Magnification: 425×

Figure 10
The development of fraternal and identical twins is different.

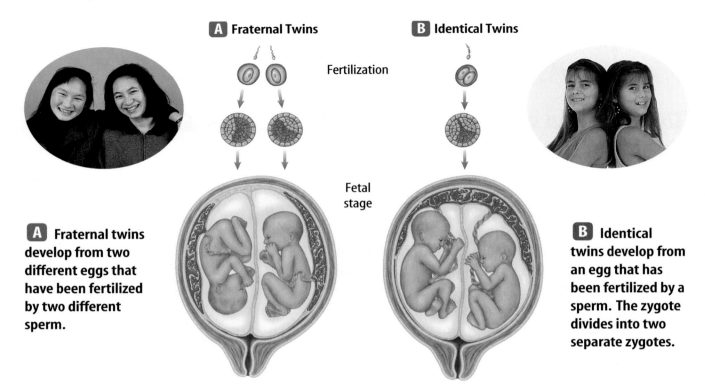

A Fraternal Twins

B Identical Twins

Fertilization

Fetal stage

A Fraternal twins develop from two different eggs that have been fertilized by two different sperm.

B Identical twins develop from an egg that has been fertilized by a sperm. The zygote divides into two separate zygotes.

Multiple Births

Sometimes two eggs leave the ovary at the same time. If both eggs are fertilized and both develop, fraternal twins are born. Fraternal twins, as shown in **Figure 10A,** can be two girls, two boys, or a boy and a girl. Because fraternal twins come from two eggs, they only resemble each other.

Because identical twin zygotes develop from the same egg and sperm, as explained in **Figure 10B,** they have the same hereditary information. These identical zygotes develop into identical twins, which are either two girls or two boys. Multiple births also can occur when three or more eggs are produced at one time or when the zygote separates into three or more parts.

Development Before Birth

After fertilization, the zygote moves along the oviduct to the uterus. During this time, the zygote is dividing and forming into a ball of cells. After about seven days, the zygote attaches to the wall of the uterus, which has been thickening in preparation to receive a zygote, as shown in **Figure 11.** If attached to the wall of the uterus, the zygote will develop into a baby in about nine months. This period of development from fertilized egg to birth is known as **pregnancy.**

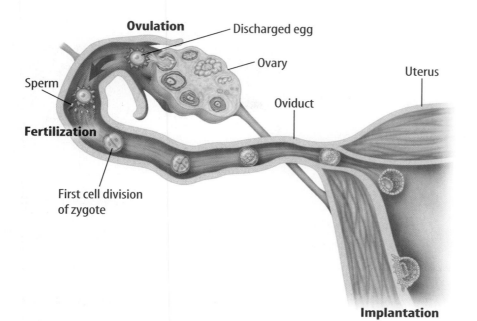

Ovulation — Discharged egg

— Ovary

Sperm

Fertilization

Oviduct

Uterus

First cell division
of zygote

Implantation

Figure 11
After a few days of rapid cell division, the zygote, now a ball of cells, reaches the lining of the uterus, where it attaches itself to the lining for development.

The Embryo After the zygote attaches to the wall of the uterus, it is known as an **embryo,** illustrated in **Figure 12.** It receives nutrients from fluids in the uterus until the placenta (pluh SENT uh) develops from tissues of the uterus and the embryo. An umbilical cord develops that connects the embryo to the placenta. In the placenta, materials diffuse between the mother's blood and the embryo's blood, but their bloods do not mix. Blood vessels in the umbilical cord carry nutrients and oxygen from the mother's blood through the placenta to the embryo. Other substances in the mother's blood can move into the embryo, including drugs, toxins, and disease organisms. Wastes from the embryo are carried in other blood vessels in the umbilical cord through the placenta to the mother's blood.

✔ **Reading Check** *Why must a pregnant woman avoid alcohol, tobacco, and harmful drugs?*

Pregnancy in humans lasts about 38 to 39 weeks. During the third week, a thin membrane called the **amniotic** (am nee AH tihk) **sac** begins to form around the embryo. The amniotic sac is filled with a clear liquid called amniotic fluid, which acts as a cushion for the embryo and stores nutrients and wastes.

During the first two months of development, the embryo's major organs form and the heart structure begins to beat. At five weeks, the embryo has a head with eyes, nose, and mouth features. During the sixth and seventh weeks, fingers and toes develop.

Figure 12
By two months, the developing embryo is about 2.5 cm long and is beginning to develop recognizable features.

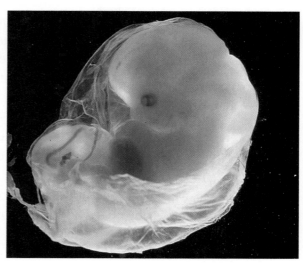

Figure 13
A fetus at about 16 weeks is approximately 15 cm long and weighs 140 g.

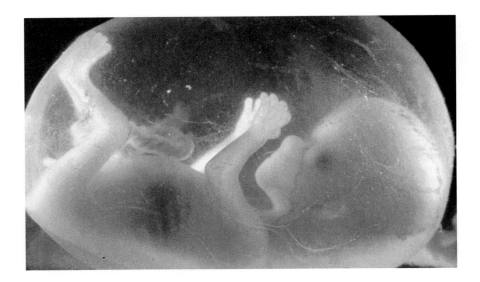

Interpreting Fetal Development

Procedure
Make a bar graph of the following data.

Fetal Development	
End of Month	Length (cm)
3	8
4	15
5	25
6	30
7	35
8	40
9	51

Analysis
1. During which month does the greatest increase in length occur?
2. On average, how many centimeters does the baby grow per month?

The Fetus After the first two months of pregnancy, the developing embryo is called a **fetus,** shown in **Figure 13.** At this time, body organs are present. Around the third month, the fetus is 8 cm to 9 cm long. The mother may feel the fetus move. The fetus can even suck its thumb. By the fourth month, an ultrasound test can determine the sex of the fetus. The fetus is 30 cm to 38 cm in length by the end of the seventh month of pregnancy. Fatty tissue builds up under the skin, and the fetus looks less wrinkled. By the ninth month, the fetus usually has shifted to a head-down position within the uterus, a position beneficial for delivery. The head usually is in contact with the opening of the uterus to the vagina. The fetus is about 50 cm in length and weighs from 2.5 kg to 3.5 kg.

The Birthing Process

The process of childbirth, as shown in **Figure 14,** begins with labor, the muscular contractions of the uterus. As the contractions increase in strength and number, the amniotic sac usually breaks and releases its fluid. Over a period of hours, the contractions cause the opening of the uterus to widen. More powerful and more frequent contractions push the baby out through the vagina into its new environment.

Delivery Often a mother is given assistance by a doctor during the delivery of the baby. As the baby emerges from the birth canal, a check is made to determine if the umbilical cord is wrapped around the baby's neck or any body part. When the head is free, any fluid in the baby's nose and mouth is removed by suction. After the head and shoulders appear, contractions force the baby out completely. Up to an hour after delivery, contractions occur that push the placenta out of the mother's body.

Cesarean Section Sometimes a baby must be delivered before labor begins or before it is completed. At other times, a baby cannot be delivered through the birth canal because the mother's pelvis might be too small or the baby might be in the wrong birthing position. In cases like these, surgery called a cesarean (suh SEER ee uhn) section is performed. An incision is made through the mother's abdominal wall, then through the wall of the uterus. The baby is delivered through this opening.

✔ **Reading Check** *What is a cesarean section?*

After Birth When the baby is born, it is attached to the umbilical cord. The person assisting with the birth clamps the cord in two places and cuts it between the clamps. The baby does not feel any pain from this procedure. The baby might cry, which is the result of air being forced into its lungs. The scar that forms where the cord was attached is called the navel.

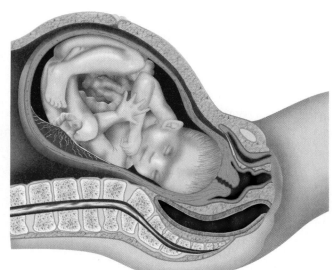

A The fetus moves into the opening of the birth canal, and the uterus begins to widen.

Figure 14
Childbirth begins with labor. The opening to the uterus widens, and the baby passes through.

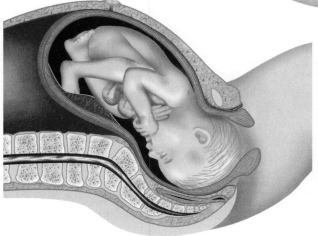

B The base of the uterus is completely dilated.

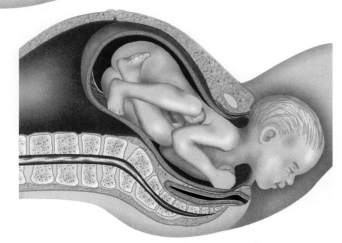

C The fetus is pushed out through the birth canal.

Stages After Birth

Defined stages of development occur after birth, based on the major developments that take place during those specific years. Infancy lasts from birth to around 18 months of age. Childhood extends from the end of infancy to sexual maturity, or puberty. The years of adolescence vary, but they usually are considered to be the teen years. Adulthood covers the years of age from the early 20s until life ends, with older adulthood considered to be over 60. The age spans of these different stages are not set, and scientists differ in their opinions regarding them.

Infancy What type of environment must the infant adjust to after birth? The experiences the fetus goes through during birth cause **fetal stress.** The fetus has emerged from an environment that was dark, watery, a constant temperature, and nearly soundless. In addition, the fetus might have been forced through the constricted birth canal. However, in a short period of time, the infant's body becomes adapted to its new world.

The first four weeks after birth are known as the neonatal(nee oh NAY tul) period. The term *neonatal* means "newborn." During this time, the baby's body begins to function normally. Unlike the newborn of some other animals, human babies, shown in **Figure 15A,** depend on other humans for their survival. In contrast, many other animals, such as horses like those shown in **Figure 15B,** begin walking a few hours after they are born.

Figure 15
Human babies are more dependent upon their caregivers than many other mammals are.

A Infants and toddlers are completely dependent upon caregivers for all their needs.

B Other young mammals are more self-sufficient. This colt is able to stand within an hour after birth.

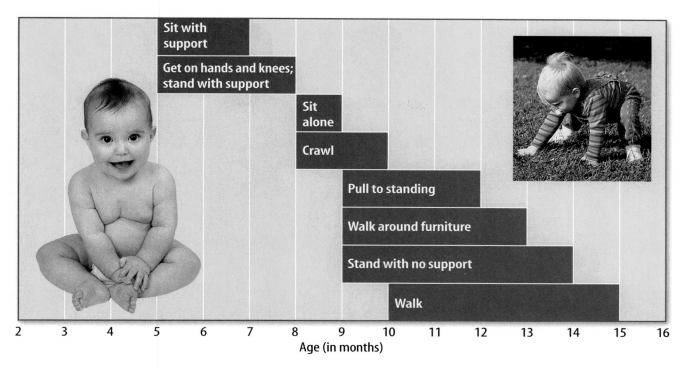

Sit with support

Get on hands and knees; stand with support

Sit alone

Crawl

Pull to standing

Walk around furniture

Stand with no support

Walk

| 2 | 3 | 4 | 5 | 6 | 7 | 8 | 9 | 10 | 11 | 12 | 13 | 14 | 15 | 16 |

Age (in months)

During these first 18 months, infants show increased physical coordination, mental development, and rapid growth. Many infants will triple their weight in the first year. **Figure 16** shows the extremely rapid development of the nervous and muscular systems during this stage, which enables infants to start interacting with the world around them.

Figure 16
Infants show rapid development in their nervous and muscular systems through 18 months of age.

Childhood After infancy is childhood, which lasts until about puberty, or sexual maturity. Sexual maturity occurs around 12 years of age. Overall, growth during early childhood is rather rapid, although the physical growth rate for height and weight is not as rapid as it is in infancy. Between two and three years of age, the child learns to control his or her bladder and bowels. At age two to three, most children can speak in simple sentences. Around age four, the child is able to get dressed and undressed with some help. By age five, many children can read a limited number of words. By age six, children usually have lost their chubby baby appearance, as seen in **Figure 17.** However, muscular coordination and mental abilities continue to develop. Throughout this stage, children develop their abilities to speak, read, write, and reason. These ages of development are only guidelines because each child develops at a different rate.

Figure 17
Children grow and develop at different rates, like these kindergartners.

Figure 18
The proportions of body parts change over time as the body develops. *Describe how the head changes proportion.*

Physics
INTEGRATION

During adolescence, the body parts do not all grow at the same rate. The legs grow longer before the upper body lengthens. This changes the body's center of gravity, the point at which the body maintains its balance. This is one cause of teenager clumsiness. In your Science Journal, write a paragraph about how this might affect playing sports.

Adolescence Adolescence usually begins around age 12 or 13. A part of adolescence is puberty—the time of development when a person becomes physically able to reproduce. For girls, puberty occurs between ages nine and 13. For boys, puberty occurs between ages 13 and 16. During puberty, hormones produced by the pituitary gland cause changes in the body. These hormones produce reproductive cells and sex hormones. Secondary sex characteristics also develop. In females, the breasts develop, pubic and underarm hair appears, and fatty tissue is added to the buttocks and thighs. In males, the hormones cause a deepened voice, an increase in muscle size, and the growth of facial, pubic, and underarm hair.

Adolescence usually is when the final growth spurt occurs. Because the time when hormones begin working varies among individuals and between males and females, growth rates differ. Girls often begin their final growth phase at about age 11 and end around age 16. Boys usually start their growth spurt at age 13 and end around 18 years of age.

Adulthood The final stage of development, adulthood, begins with the end of adolescence and continues through old age. This is when the growth of the muscular and skeletal system stops. **Figure 18** shows how body proportions change as you age.

People from age 45 to age 60 are sometimes considered middle-aged adults. During these years, physical strength begins to decline. Blood circulation and respiration become less efficient. Bones become more brittle, and the skin becomes wrinkled.

Older Adulthood People over the age of 60 may experience an overall decline in their physical body systems. The cells that make up these systems no longer function as well as they did at a younger age. Connective tissues lose their elasticity, causing muscles and joints to be less flexible. Bones become thinner and more brittle. Hearing and vision are less sensitive. The lungs and heart work less efficiently. However, exercise and eating well over a lifetime can help extend the health of one's body systems. Many healthy older adults enjoy full lives and embrace challenges, as shown in **Figure 19.**

Figure 19
Astronaut and Senator John Glenn traveled into space twice. In 1962, at age 40, he was the first U.S. citizen to orbit Earth. He was part of the space shuttle crew in 1998 at age 77. Senator Glenn has helped change people's views of what many older adults are capable of doing.

 **Reading Check** *What physical changes occur during late adulthood?*

Human Life Spans Seventy-five years is the average life span—from birth to death—of humans, although an increasing number of people live much longer. However, body systems break down with age, resulting in eventual death. Death can occur earlier than old age for many reasons, including diseases, accidents, and bad health choices.

Section Assessment

1. What happens when an egg is fertilized in a female?

2. What happens to an embryo during the first two months of pregnancy?

3. Describe the major events that occur during childbirth.

4. What stage of development are you in? What physical changes have occurred or will occur during this stage of human development?

5. **Think Critically** Why is it hard to compare the growth and development of different adolescents?

Skill Builder Activities

6. **Making Models** Use references to construct a time line that highlights the major events in the various stages of development from the embryo to adulthood. **For more help, refer to the** Science Skill Handbook.

7. **Using an Electronic Spreadsheet** Using your text and other resources, make a spreadsheet for the stages of human development from a zygote to a fetus. Title one column *Zygote,* another *Embryo,* and a third *Fetus.* Complete the spreadsheet. **For more help, refer to the** Technology Skill Handbook.

Activity

Changing Body Proportions

The ancient Greeks believed the perfect body was completely balanced. Arms and legs should not be too long or short. A person's head should not be too large or small. The extra large muscles of a body builder would have been ugly to the Greeks. How do you think they viewed the bodies of infants and children? Infants and young children have much different body proportions than adults, and teenagers often go through growth spurts that quickly change their body proportions. How do body proportions differ among people?

What You'll Investigate

How do the body proportions differ between adolescent males and females?

Materials
tape measure
erasable pencil
graph paper

Goals
- **Measure** specific body proportions of adolescents.
- **Infer** how body proportions differ between adolescent males and females.

Procedure

1. Copy the data table in your Science Journal and record the gender of each person that you measure.

2. Measure each person's head circumference by starting in the middle of the forehead and wrapping the tape measure around the head. Record these measurements.

3. Measure each person's arm length from the top of the shoulder to the tip of the middle finger while the arm is held straight out to the side of the body. Record these measurements.

4. Ask each person to remove his or her shoes and stand next to a wall. Mark their height with an erasable pencil and measure their height from the floor to the mark. Record these measurements in the data table.

5. **Combine** your data with that of your classmates. Find the averages of head circumference, arm length, and height. Then, find these averages for males and females.

6. Make a bar graph of your calculations in step 5. Plot the measurements on the *y*-axis and plot all of the averages along the *x*-axis.

7. **Calculate** the proportion of average head circumference to average height for everyone in your class by dividing the average head circumference by the average height. Repeat this calculation for males and females.

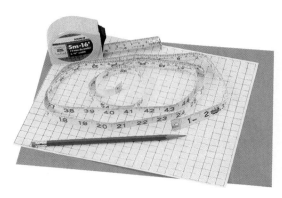

8. **Calculate** the proportion of average arm length to average height for everyone in your class by dividing the average arm length by the average height. Repeat this calculation for males and females.

Age and Body Measurements			
Gender of Person	Head Circumference (cm)	Arm Length (cm)	Height (cm)

Conclude and Apply

1. Do adolescent males or females have larger head circumferences or longer arms? Which group has the larger proportion of head circumference or arm length to height?

2. Does this activity support the information in this chapter about the differences between growth rates of adolescent males and females? Explain.

*C*ommunicating Your Data

On poster board, **construct** data tables showing your results and those of your classmates. Discuss with your classmates why these results might be different.

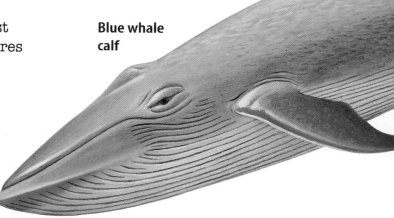

Science Stats

Facts About Infants

Did you know...

...Humans and chimpanzees share about 99 percent of their genes. Although humans look different than chimps, reproduction is similar and gestation is the same—about nine months. Youngsters of both species lose their baby teeth at about six years of age.

Female kangaroo and joey

...Unlike humans and most other mammals, the newborn kangaroo develops in its mother's pouch longer than in her uterus. About one month after fertilization, the kangaroo is born and moves from its mother's uterus into her pouch. It spends seven to ten months in the pouch, drinking milk and growing.

...The blue whale calf is the biggest newborn in the world. At birth, it measures about 7 m and weighs about 2,700 kg. The average human newborn measures about 50 cm and weighs about 3.3 kg. In its first year of life, a blue whale calf gains about 90 kg every day. Compare that to an average human baby who gains about 10 kg during his or her entire first year.

Blue whale calf

Mammal Facts

Mammal	Average Gestation	Average Birth Weight	Average Adult Weight	Average Life Span (years)
African Elephant	22 months	136 kg	4,989.5 kg	35
Blue Whale	12 months	1,800 kg	135,000 kg	60
Human	**9 months**	**3.3 kg**	**59–76 kg**	**76***
Brown Bear	7 months	0.23–0.5 kg	350 kg	22.5
Cat	2 months	99 g	2.7–7 kg	13.5
Kangaroo	1 month	0.75–1.0 g	45 kg	5
Golden Hamster	2.5 weeks	0.3 g	112 g	2

* In the United States

1-day-old to 7-day-old mice

...Of about 4,000 species of mammals, only three lay eggs. These species are the platypus, the short-beaked echidna (ih KIHD nuh), and the long-beaked echidna. No other mammals lay eggs.

Echidna

... House mice can have up to ten litters per year, each one containing up to seven mice. Mice have this many offspring because so few of them survive.

Do the Math

1. Look at the data table above. Make a generalization about a mammal's birth weight and the length of its gestation period.
2. Make a bar graph that compares the length of the human gestation period to that of two other mammals.
3. Assume that a female of each mammal listed in the table above is pregnant once during her life. Which mammal is pregnant for the greatest proportion of her life?

Go Further

Do research to find out which species of vertebrate animals has the longest life span and which has the shortest. Present your findings in a table that also shows the life span of humans.

Chapter 6 Study Guide

Reviewing Main Ideas

Section 1 The Endocrine System

1. Endocrine glands secrete hormones directly into the bloodstream.

2. Hormones affect specific tissues throughout the body. *How can a gland near your head control chemical activities in other parts of your body?*

3. A change in the body causes an endocrine gland to function. When homeostasis is reached, the endocrine gland receives a signal to slow or stop its production.

Section 2 The Reproductive System

1. The reproductive system allows new organisms to be formed.

2. The testes produce sperm that leave the male through the penis.

3. The female ovary produces an egg. If fertilized, it becomes a zygote and later develops into a fetus within the uterus. *How are the structures of the egg and sperm suited for their functions?*

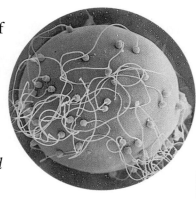

4. When an egg is not fertilized, the built-up lining of the uterus is shed in a process called menstruation. This process begins 14 days after ovulation.

Section 3 Human Life Stages

1. After fertilization, the zygote undergoes developmental changes to become an embryo, then a fetus. Twins occur when two eggs are fertilized or when a zygote divides after fertilization.

2. Birth begins with labor—muscular contractions of the uterus. The amniotic sac breaks. Then, usually after several hours, the contractions force the baby out of the mother's body. *Why is the first stage in the birthing process called labor?*

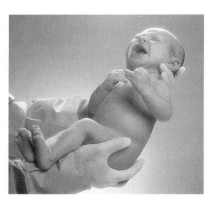

3. Infancy is the stage of development from birth to 18 months of age. It is a period of rapid growth of mental and physical skills. Childhood, which lasts until age 12, is marked by development of muscular coordination and mental abilities.

4. Adolescence is the stage of development when a person becomes physically able to reproduce. The final stage of development is adulthood. Physical development is complete and body systems become less efficient. Death occurs at the end of life.

FOLDABLES Reading & Study Skills

After You Read

Using the information on your Foldable as an outline, explain each stage of human life.

Visualizing Main Ideas

Complete the following table on life stages.

Human Development

Stages of Life	Age Range	Physical Development
Infant		sits, stands, words spoken
		walks, speaks, writes, reads
Adolescent		
		end of muscular and skeletal growth

Vocabulary Review

Vocabulary Words

a. amniotic sac
b. embryo
c. fetal stress
d. fetus
e. hormone
f. menstrual cycle
g. menstruation
h. ovary

i. ovulation
j. pregnancy
k. semen
l. sperm
m. testes
n. uterus
o. vagina

THE PRINCETON REVIEW **Study Tip**

Use lists to help you memorize facts. For example, when trying to memorize the stages of human development, write them down several times on a piece of paper until you know them.

Using Vocabulary

Replace the underlined words with the correct vocabulary word(s).

1. Testes is a mixture of sperm and fluid.

2. The time of the development until the birth of a baby is known as menstruation.

3. During the first two months of pregnancy, the unborn child is known as fetal stress.

4. The vagina is a hollow, pear-shaped muscular organ.

5. The ovary is the membrane that protects the unborn child.

6. After two months of pregnancy, the unborn child is known as a(n) embryo.

7. The testes is the organ that produces eggs.

Checking Concepts

Choose the word or phrase that best answers the question.

1. What are the chemicals produced by the endocrine system?
   A) enzymes
   C) hormones
   B) target tissues
   D) saliva

2. Which gland produces melatonin?
   A) adrenal
   C) pancreas
   B) thyroid
   D) pineal

3. Where does the embryo develop?
   A) oviduct
   C) uterus
   B) ovary
   D) vagina

4. What is the monthly process that releases an egg called?
   A) fertilization
   C) menstruation
   B) ovulation
   D) puberty

5. What is the union of an egg and a sperm?
   A) fertilization
   C) menstruation
   B) ovulation
   D) puberty

6. Where is the egg usually fertilized?
   A) oviduct
   C) vagina
   B) uterus
   D) ovary

7. When does puberty occur?
   A) childhood
   C) adolescence
   B) adulthood
   D) infancy

8. Which sex characteristics are common to males and females?
   A) breasts
   C) increased fat
   B) pubic hair
   D) increased muscles

9. During which period does growth stop?
   A) childhood
   C) adolescence
   B) adulthood
   D) infancy

10. During what stage of development does the amniotic sac form?
    A) zygote
    C) fetus
    B) embryo
    D) newborn

Thinking Critically

11. List the effects that adrenal gland hormones can have on your body as you prepare to run a race.

12. Explain the similar functions of the ovaries and testes.

13. Identify the structure in the following diagram in which each process occurs: ovulation, fertilization, and implantation.

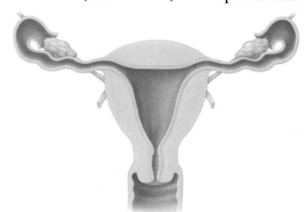

14. How is your endocrine system like the thermostat in your house?

15. Are quadruplets always identical or always fraternal, or can they be either? Explain.

Developing Skills

16. **Predicting** During the ninth month of pregnancy, the fetus develops a white, greasy coating. Predict what the function of this coating might be.

17. **Forming Hypotheses** Make a hypothesis about the effect of raising identical twins apart from each other.

18. **Classifying** Classify each of the following structures as female or male and internal or external: ovary, penis, scrotum, testes, uterus, and vagina.

19. Concept Mapping Complete the following concept map of egg release and implantation using the appropriate scientific words.

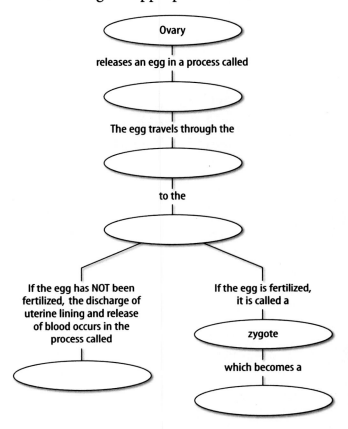

=== **Performance Assessment** ===

20. Letter Find newspaper or magazine articles on the effects of smoking on the health of the developing embryo and newborn. Write a letter to the editor about why a mother's smoking is damaging her unborn baby's health.

TECHNOLOGY

Go to the Glencoe Science Web site at **science.glencoe.com** or use the **Glencoe Science CD-ROM** for additional chapter assessment.

 **Test Practice**

In health class, Angela decided to do a report about the cases of syphilis in the United States. She brought the following graph to accompany her report, which shows syphilis rates by year between 1970 and 1997.

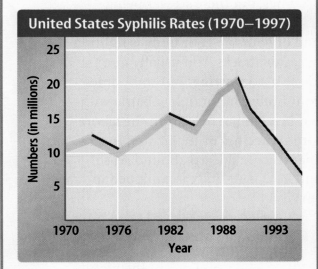

Study the graph and answer the following questions.

1. According to the information in the graph, when did an epidemic of syphilis occur in the United States?
   A) 1970–1972
   B) 1982–1984
   C) 1988–1990
   D) 1992–1994

2. A reasonable hypothesis for the information in the graph is that the number of people infected with syphilis _____.
   F) is increasing
   G) is decreasing
   H) has remained the same
   J) is related to gender

Immunity and Disease

It never makes newspaper headlines, but there's a war being fought in your body. Every second of your life your body is fighting harmful attacks. You usually don't know it's occurring. But sometimes your body cannot fight a battle without bringing in help from the laboratory—vaccines or medicines. In this chapter, you'll learn about disease and how your body is equipped to survive.

What do you think?

Science Journal Look at the picture below with a classmate. Discuss what this might be. Here's a hint: *You can receive a booster shot so you don't get this disease from dirty cuts.* Write your answer in your Science Journal.

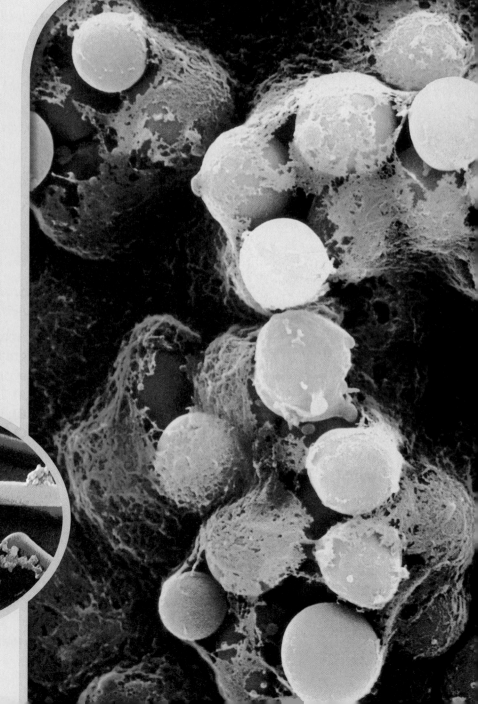

EXPLORE **A**CTIVITY

It is a fact that disease-causing organisms are in the air you breathe and on the objects you touch. Knowing how diseases are spread will help you understand how your body fights disease. You can discover one way diseases are spread by doing the following activity.

Model the spread of disease-causing organisms

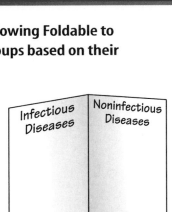

1. Wash your hands before and after this activity. Don't touch your face until the activity is completed and your hands are washed.
2. Work with a partner. Place a drop of peppermint food flavoring on a cotton ball. Pretend that the flavoring is a mass of cold viruses.
3. Use the cotton ball to rub an X over the palm of your right hand. Let it dry.
4. Shake hands with your partner.
5. Have your partner shake hands with another student. Then each student should smell their hands

Observe

Observe how many persons your "virus" infected. Describe in your Science Journal some ways diseases are spread. How could some of these diseases be stopped?

FOLDABLES
Reading & Study
Skills

Before You Read

Making a Classify Study Fold Make the following Foldable to help organize objects or events into their groups based on their common features.

1. Place a sheet of paper in front of you so the long side is at the top. Fold the paper in half from the left side to the right side. Then unfold.
2. Label the left side of the paper *Infectious Diseases* and the right side of the paper *Noninfectious Diseases* as shown.
3. Before you read the chapter, classify diseases you are familiar with as infectious or noninfectious by listing them on the proper fold.
4. As you read the chapter, change and add to your lists.

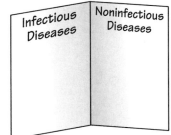

Infectious Diseases | Noninfectious Diseases

The Immune System

As You Read

What You'll Learn

- **Describe** the natural defenses your body has against disease.
- **Explain** the difference between an antigen and an antibody.
- **Compare and contrast** active and passive immunity.

Vocabulary

immune system   active immunity
antigen   passive immunity
antibody   vaccination

Why It's Important

Your body's defenses fight the pathogens that you are exposed to every day.

Lines of Defense

The Sun has just begun to peek over the horizon, casting an orange glow on the land. A skunk ambles down a dirt path. Behind the skunk, you and your dog come over a hill for your morning exercise. Suddenly, the skunk stops and raises its tail high in the air. Your dog creeps forward. "No!" you shout. The dog ignores your command. Without further warning, the skunk sprays your dog. Yelping pitifully and carrying an awful stench, your dog takes off. The skunk used its scent to protect itself. Its first-line defense was to warn your dog with its posture. Its second-line defense was its spray. Just as the skunk protects itself from predators, your body also protects itself from harm.

Your body has many ways to defend itself. Its first-line defenses work against harmful substances and all types of disease-causing organisms, called pathogens (PA thuh junz). Your second-line defenses are specific and work against specific pathogens. This complex group of defenses is called your **immune system.** Tonsils, shown in **Figure 1,** are one of the immune system organs that protect your body.

Reading Check   *What types of defenses does your body have?*

First-Line Defenses Your skin and respiratory, digestive, and circulatory systems are first-line defenses against pathogens. Skin is a barrier that prevents many pathogens from entering your body. Although most pathogens can't get through unbroken skin, as shown in **Figure 2,** they can get into your body easily through a cut or through your mouth and the membranes in your nose and eyes. The conditions on the skin can affect pathogens. Perspiration contains substances that can slow the growth of some pathogens. At times, secretions from the skin's oil glands and perspiration are acidic. Some pathogens cannot grow in this acidic environment.

Figure 1
Tonsils help prevent infection in your respiratory and digestive tract.

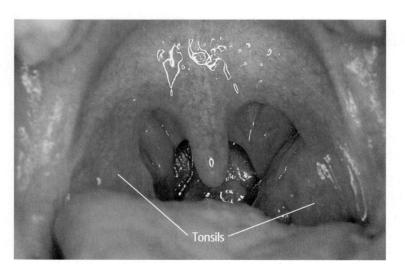

Tonsils

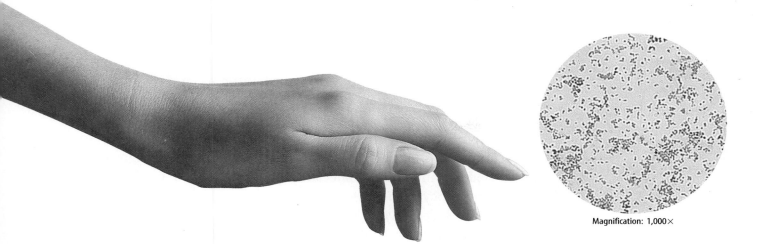

Magnification: 1,000×

Internal First-Line Defenses Your respiratory system traps pathogens with hairlike structures, called cilia (SIH lee uh), and mucus. Mucus contains an enzyme that weakens the cell walls of some pathogens. When you cough or sneeze, you get rid of some of these trapped pathogens.

Your digestive system has several defenses against pathogens—saliva, enzymes, hydrochloric acid, and mucus. Saliva in your mouth contains substances that kill bacteria. Also, enzymes (EN zimez) in your stomach, pancreas, and liver help destroy pathogens. Hydrochloric acid in your stomach helps digest your food. It also kills some bacteria and stops the activity of some viruses that enter your body on the food that you eat. The mucus found on the walls of your digestive tract contains a chemical that coats bacteria and prevents them from binding to the inner lining of your digestive organs.

Your circulatory system contains white blood cells, like the one in **Figure 3,** that surround and digest foreign organisms and chemicals. These white blood cells constantly patrol your body, sweeping up and digesting bacteria that invade. They slip between cells of tiny blood vessels called capillaries. If the white blood cells cannot destroy the bacteria fast enough, you might develop a fever. Many pathogens are sensitive to temperature. A slight increase in body temperature slows their growth and activity but speeds up your body's defenses.

Inflammation When tissue is damaged by injury or infected by pathogens, it becomes inflamed. Signs of inflammation include redness, temperature increase, swelling, and pain. Chemical substances released by damaged cells cause capillary walls to expand, allowing more blood to flow into the area. Other chemicals released by damaged tissue attract certain white blood cells that surround and take in pathogenic bacteria. If pathogens get past these first-line defenses, your body uses another line of defense called specific immunity.

Figure 2
Most pathogens, like the staphylococci bacteria shown here, cannot get through un-broken skin.

Figure 3
A white blood cell leaves a capillary. It will search out and destroy harmful microorganisms in your body tissues.

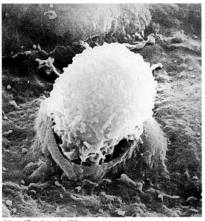

Magnification: 3,450×

Specific Immunity When your body fights disease, it is battling complex molecules that don't belong there. Molecules that are foreign to your body are called **antigens** (AN tih junz). Antigens can be separate molecules or they can be found on the surface of a pathogen. For example, the protein in the cell membrane of a bacterium can be an antigen. When your immune system recognizes molecules as being foreign to your body, as in **Figure 4,** special lymphocytes called T cells respond. Lymphocytes are a type of white blood cell. One type of T cells, called killer T cells, releases enzymes that help destroy invading foreign matter. Another type of T cells, called helper T cells, turns on the immune system. They stimulate other lymphocytes, known as B cells, to form antibodies.

An **antibody** is a protein made in response to a specific antigen. The antibody attaches to the antigen and makes it useless. This can happen in several ways. The pathogen might not be able to stay attached to a cell. It might be changed in such a way that a killer T cell can capture it more easily or the pathogen can be destroyed.

Reading Check *What is an antibody?*

Another type of lymphocyte, called memory B cells, also has antibodies for the specific pathogen. Memory B cells remain in the blood ready to defend against an invasion by that same pathogen another time.

Figure 4
The response of your immune system to disease-causing organisms can be divided into four steps—recognition, mobilization, disposal, and immunity. *What is the function of B cells?*

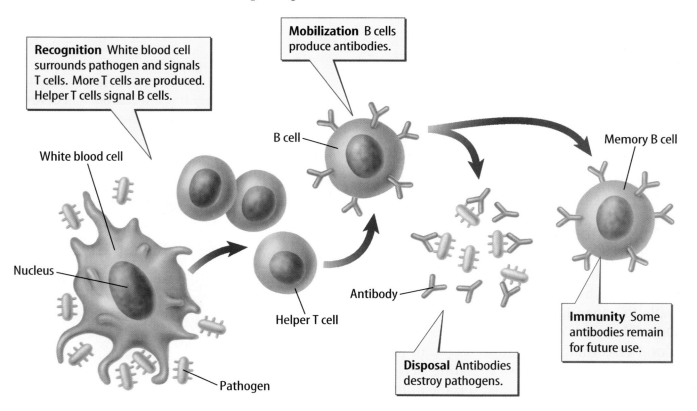

Recognition White blood cell surrounds pathogen and signals T cells. More T cells are produced. Helper T cells signal B cells.

Mobilization B cells produce antibodies.

Disposal Antibodies destroy pathogens.

Immunity Some antibodies remain for future use.

White blood cell · Nucleus · Pathogen · B cell · Helper T cell · Antibody · Memory B cell

Active Immunity Antibodies help your body build defenses in two ways—actively and passively. In **active immunity** your body makes its own antibodies in response to an antigen. **Passive immunity** results when antibodies that have been produced in another animal are introduced into your body.

When a pathogen invades your body, the pathogen quickly multiplies and you get sick. Your body immediately starts to make antibodies to attack the pathogen. After enough antibodies form, you usually get better. Some antibodies stay on duty in your blood, and more are produced rapidly if the pathogen enters your body again. Because of this defense system you usually don't get certain diseases such as chicken pox more than once.

Vaccination Another way to develop active immunity to a disease is to be inoculated with a vaccine. The process of giving a vaccine by injection or by mouth is called **vaccination.** A vaccine is a form of the antigen that gives you active immunity against a disease. For example, suppose a measles vaccine is injected into your body. Your body forms antibodies against the measles antigen. If you later encounter the same measles virus, antibodies that are needed to fight and destroy the measles virus already are in your bloodstream. Vaccines have helped reduce cases of childhood diseases, as shown in **Table 1.**

Antibodies that immunize you against one virus may not guard against a different virus. For example, flu shots are given annually because each year a different flu virus causes the disease. A vaccine can prevent a disease, but it is not a cure. As you grow older, you will be exposed to many more types of pathogens and will build a separate immunity to each one.

TRY AT HOME Mini LAB

Determining Reproduction Rates

Procedure

1. Place one penny on a table. Imagine that the penny is a bacterium that can divide every 10 min.
2. Place **two pennies** below (but not under) the first penny to indicate the two bacteria present after the first bacterium divides.
3. Repeat three more divisions, placing two pennies under each penny in the row above.
4. Stop using pennies, wash your hands, and calculate how many bacteria you would have after 5 h of reproduction. Graph your data.

Analysis

1. How many bacteria are present after 5 h?
2. Why is it important to take antibiotics promptly if you have an infection?

Table 1 Cases of Disease Before and After Vaccine Availability in the U.S.		
Disease	**Average Number of Cases per Year Before Vaccine Available**	**Cases in 1998 After Vaccine Available**
Measles	503,282	89
Diptheria	175,885	1
Tetanus	1,314	34
Mumps	152,209	606
Rubella	47,745	345
Pertussis (whooping cough)	147,271	6,279

Data from the National Immunization Program, CDC

Figure 5
Between ages 14 and 16, immunization against diptheria and tetanus (DT) called booster shots are given.

Passive Immunity Passive immunity does not last as long as active immunity does. For example, you were born with all the antibodies that your mother had in her blood. However, these antibodies stayed with you for only a few months. Because newborn babies lose their passive immunity in a few months, they need to be vaccinated to develop their own immunity.

Tetanus Tetanus is a disease caused by a common soil bacterium. The bacterium produces a chemical that paralyzes muscles. Puncture wounds, deep cuts, and other wounds can be infected by this bacterium. Several times in early childhood you received active vaccines that stimulated antibody production to tetanus toxin. As shown in **Figure 5,** you should continue to get vaccines or boosters every ten years to maintain protection. Booster shots for diphtheria, which is a dangerous infectious respiratory disease, are given in the same vaccine with tetanus.

Suppose a person who hasn't been vaccinated against tetanus toxin gets a puncture wound. The person would be given passive immunity—antibodies to the toxin. These antibodies usually are from humans but can be from horses or cattle if human antibodies are not available. This passive immunity against tetanus lasts long enough to prevent the person from getting the disease. But he or she still needs to receive active vaccine to develop antibodies against tetanus and to maintain protection.

Section Assessment

1. Describe how harmful bacteria cause infections in your body.
2. List natural defenses that your body has against disease.
3. How does an active vaccine work in the human body?
4. How does your immune system react when it detects an antigen?
5. **Think Critically** Several diseases have symptoms similar to those of measles. Why doesn't the measles vaccine protect you from all of these diseases?

Skill Builder Activities

6. **Making Models** Create models of the different types of T cells, antigens, and B cells from clay, construction paper, or other art materials. Use them to explain how T cells function in the immune system. **For more help, refer to the** Science Skill Handbook.

7. **Using a Word Processor** Using the information in this section, create a flowchart that compares active immunity and passive immunity. **For more help, refer to the** Technology Skill Handbook.

Infectious Diseases

Disease in History

For thousands of years, people have feared outbreaks of disease. The plague, smallpox, and influenza have killed millions of people worldwide. It is estimated that during the 1918 influenza outbreak, 20 million to 40 million people died, as **Table 2** shows. Today, the causes of these diseases are known, and treatments can prevent or cure them. But even today, there are diseases such as the Ebola virus in Africa that cannot be cured.

Discovering Disease Organisms With the invention of the microscope in the latter part of the seventeenth century, bacteria, yeast, and mold spores were seen for the first time. However, it took almost 200 years more to discover the relationship between some of them and disease. Scientists gradually learned that microorganisms were responsible for fermentation and decay. If decay-causing microorganisms could cause changes in other organisms, it was hypothesized that microorganisms could cause diseases and carry them from one person to another. Scientists did not make a connection between viruses and disease transmission until the late 1800s and early 1900s.

The French chemist Louis Pasteur learned that microorganisms might cause disease in humans. Many scientists of his time did not believe that microorganisms could harm larger organisms, such as humans. However, Pasteur discovered that microorganisms could spoil wine and milk. He then realized that microorganisms could attack the human body in the same way. Pasteur invented **pasteurization** (pas chuh ruh ZAY shun), which is the process of heating a liquid to a specific temperature that kills most bacteria.

As You Read

What You'll Learn

■ **Describe** the work of Pasteur, Koch, and Lister in the discovery and prevention of disease.
■ **Identify** diseases caused by viruses and bacteria.
■ **List** sexually transmitted diseases, their causes, and treatments.
■ **Explain** how HIV affects the immune system.

Vocabulary
pasteurization
virus
infectious disease
biological vector
sexually transmitted disease (STD)

Why It's Important
You can help prevent certain illnesses if you know what causes disease and how disease spreads.

Table 2 Deaths from the 1918 Influenza Epidemic	
Country or Region	**Estimated Number of Deaths**
United States	550,000
United Kingdom	228,000
India	12,500,000
Africa	8,450,000
Australia and Samoa	2,988,000
Total Worldwide	20,000,000 to 40,000,000

Table 3 Human Diseases and Their Agents	
Agent	**Diseases**
Bacteria	Tetanus, tuberculosis, typhoid fever, strep throat, bacterial pneumonia, plague
Protists	Malaria, sleeping sickness
Fungi	Athlete's foot, ringworm
Viruses	Colds, influenza, AIDS, measles, mumps, polio, smallpox

Disease Organisms Today, it is known that many diseases are caused by bacteria, certain viruses, protists (PROH tihsts), or fungi. **Table 3** lists some of the diseases caused by various groups of pathogens. Many harmful bacteria that infect your body can reproduce rapidly. The conditions in your body, such as temperature and available nutrients, help the bacteria grow and multiply. Bacteria can slow down the normal growth and metabolic activities of body cells and tissues. Some bacteria even produce toxins that kill cells on contact.

A **virus** is a minute piece of genetic material surrounded by a protein coating that infects and multiplies in host cells. The host cells die when the viruses break out of them. These new viruses infect other cells, leading to the destruction of tissues or the interruption of vital body activities.

☑ Reading Check *What is the relationship between a virus and a host cell?*

Pathogenic protists, such as the organisms that cause malaria, can destroy tissues and blood cells or interfere with normal body functions. In a similar manner, fungus infections can cause athlete's foot, nonhealing wounds, chronic lung disease, or inflammation of the membranes of the brain.

Koch's Rules Many diseases caused by pathogens can be treated with medicines. In many cases, these organisms need to be identified before specific treatment can begin. Today, a method developed in the nineteenth century still is used to identify organisms.

Pasteur may have shown that bacteria cause disease, but he didn't know how to tell which specific organism causes which disease. It was a young German doctor, Robert Koch, who first developed a way to isolate and grow one type of bacterium at a time, as shown in **Figure 6.**

Earth Science
INTEGRATION

Soil contains many microorganisms—some that are harmful, such as tetanus bacteria, and some that are helpful. Some infections are treated with antibiotics made from bacteria and molds found in the soil. One such antibiotic is streptomycin. In your Science Journal, write a brief report about the drug streptomycin.

Figure 6

In the 1880s, German doctor Robert Koch developed a series of methods for identifying which organism was the cause of a particular disease. Koch's Rules are still in use today. Developed mainly for determining the cause of particular diseases in humans and other animals, these rules have been used for identifying diseases in plants as well.

Anthrax bacteria

A In every case of a particular disease, the organism thought to cause the disease—the pathogen—must be present.

B The suspected pathogen must be separated from all other organisms and grown on agar gel with no other organisms present.

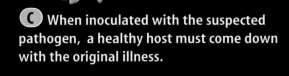

C When inoculated with the suspected pathogen, a healthy host must come down with the original illness.

Anthrax bacteria

D Finally, when the suspected pathogen is removed from the host and grown on agar gel again, it must be compared with the original organism. Only when they match can that organism be identified as the pathogen that causes the disease.

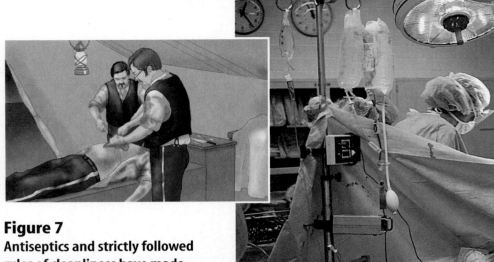

Figure 7
Antiseptics and strictly followed rules of cleanliness have made surgical procedures safer than they once were. *What differences do you see in the two operating scenes shown?*

Keeping Clean Washing your hands before or after certain activities should be part of your daily routine. Restaurant employees are required to wash their hands immediately after using the rest room. Medical professionals wash their hands before examining each patient. However, hand washing was not always a routine, even for doctors. Into the late 1800s, doctors such as those in **Figure 7** regularly operated in their street clothes and with bare, unwashed hands. A bloody apron and well-used tools were considered signs of prestige for a surgeon. More patients died from the infections that they contracted during or after the surgery than from the surgery itself.

Joseph Lister, an English surgeon, recognized the relationship between the infection rate and cleanliness. Lister dramatically reduced the number of deaths among his patients by washing their skin and his hands with carbolic (kar BAH lihk) acid, which is a liquid that kills pathogens. Lister also used carbolic acid to clean his instruments and soak bandages, and he even sprayed the air with it. The odor was strong and it irritated the skin, but more and more people began to survive surgical procedures.

Modern Operating Procedures Today antiseptics and antiseptic soaps are used to kill pathogens on skin. Every person on the surgical team washes his or her hands thoroughly and wears sterile gloves and a covering gown. The patient's skin is cleaned around the area of the body to be operated on and then covered with sterile cloths. Tools that are used to operate on the patient and all operating room equipment also are sterilized. Even the air is filtered.

 **Reading Check** *What are three ways that pathogens are reduced in today's operating room?*

How Diseases Are Spread

You walk into your kitchen before school. Your younger sister sits at the table eating a bowl of cereal. She has a fever, a runny nose, and a cough. She coughs loudly. "Hey, cover your mouth! I don't want to catch your cold," you tell her. A disease that is caused by a virus, bacterium, protist, or fungus and is spread from an infected organism or the environment to another organism is called an **infectious disease.** Infectious diseases are spread by direct contact with the infected organism, through water and air, on food, by contact with contaminated objects, and by disease-carrying organisms called **biological vectors.** Examples of vectors that have been sources of disease are rats, birds, cats, dogs, mosquitoes, fleas, and flies, as shown in **Figure 8.**

People also can be carriers of disease. When you have influenza and sneeze, you expel thousands of virus particles into the air. Colds and many other diseases are spread through contact. Each time you turn a doorknob, press the button on a water fountain, or use a telephone, your skin comes in contact with bacteria and viruses, which is why regular handwashing is recommended. The Centers for Disease Control and Prevention (CDC) in Atlanta, Georgia, monitors the spread of diseases throughout the United States. The CDC also tracks worldwide epidemics and watches for diseases brought into the United States.

Figure 8
When flies land on food, they can transport pathogens from one person to another.

Problem-Solving Activity

Has the annual percentage of deaths from major diseases changed?

Each year, many people die from diseases. Medical science has found numerous ways to treat and cure disease. Have new medicines, improved surgery techniques, and healthier lifestyles helped decrease the number of deaths from disease? By using your ability to interpret data tables, you can find out.

Identifying the Problem

The table to the right shows the percentage of total deaths due to six major diseases for a 45-year time period. Study the data for each disease. Can you see any trends in the percentage of deaths?

Solving the Problem
1. Has the percentage increased for any disease that is listed?
2. What factors could have contributed to this increase?

Percentage of Deaths Due to Major Diseases				
Disease	**Year**			
	1950	1980	1990	1995
Heart	37.1	38.3	33.5	32.0
Cancer	14.6	20.9	23.5	23.3
Stroke	10.8	8.6	6.7	6.8
Diabetes	1.7	1.8	2.2	2.6
Pneumonia and Flu	3.3	2.7	3.7	3.6
Tuberculosis	2.3	0.1	0.08	0.06

Sexually Transmitted Diseases

Infectious diseases that are passed from person to person during sexual contact are called **sexually transmitted diseases (STDs)**. STDs are caused by bacteria or viruses.

Bacterial STDs Gonorrhea (gah nuh REE uh), and chlamydia (kluh MIH dee uh) are STDs caused by bacteria. The bacteria that cause gonorrhea are shown in **Figure 9A.** A person may have one of these diseases for some time before symptoms appear. When symptoms do appear, they can include painful urination, genital discharge, and genital sores. Antibiotics are used to treat these diseases. Some of the bacteria that cause gonorrhea may be resistant to the antibiotics usually used to treat the infection. However, the disease usually can be treated with other antibiotics. If they are untreated, gonorrhea and chlamydia can leave a person sterile because the reproductive organs can be damaged permanently.

The spiral-shaped bacterium that causes syphilis (SIH fuh lus) is seen in **Figure 9B.** Syphilis has three stages. In stage 1, a sore that lasts 10 to 14 days appears on the mouth or genitals. Stage 2 may involve a rash, fever, and swollen lymph glands. Within weeks to a year, these symptoms usually disappear. The person with syphilis often believes that the disease has gone away, but it hasn't. If he or she does not seek treatment, the disease advances to stage 3, when syphilis may infect the cardiovascular and nervous systems. In all stages, syphilis is treatable with antibiotics. However, the damage to body organs in stage 3 cannot be reversed and death can result.

Viral STDs Genital herpes, a lifelong viral disease, causes painful blisters on the sex organs. This type of herpes can be transmitted during sexual contact or from an infected mother to her child during birth. The herpes virus hides in the body for long periods of time and then reappears suddenly. Herpes has no cure, and no vaccine can prevent it. However, the symptoms of herpes can be treated with antiviral medicines.

Figure 9
Bacteria that cause gonorrhea and syphilis can be destroyed with antibiotics.

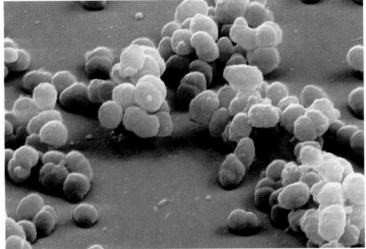

Magnification: 1,200×

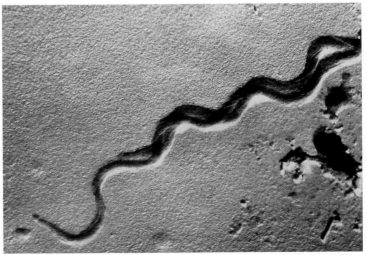

Magnification: 19,000×

HIV and Your Immune System

Human immunodeficiency virus (HIV) can exist in blood and body fluids. This virus can hide in body cells, sometimes for years. You can become infected with HIV by having sex with an HIV-infected person or by reusing an HIV-contaminated hypodermic needle for an injection. However, a freshly unwrapped sterile needle cannot transmit infection. The risk of getting HIV through blood transfusion is small because all donated blood is tested for the presence of HIV. A pregnant female with HIV can infect her child when the virus passes through the placenta. The child also may become infected from contacts with blood during the birth process or when nursing after birth.

 **Reading Check** *What are ways that a person can become infected with HIV?*

HIV cannot multiply outside the body, and it does not survive long in the environment. The virus cannot be transmitted by touching an infected person, by handling objects used by the person unless they are contaminated with body fluids, or from contact with a toilet seat.

AIDS An HIV infection can lead to Acquired Immune Deficiency Syndrome (AIDS), which is a disease that attacks the body's immune system. HIV, as shown in **Figure 10,** is different from other viruses. It attacks the helper T cells in the immune system. The virus enters the T cell and multiplies. When the infected cell bursts open, it releases more HIV. These infect other T cells. Soon, so many T cells are destroyed that not enough B cells are stimulated to produce antibodies. The body no longer has an effective way to fight invading antigens. The immune system then is unable to fight HIV or any other pathogen. For this reason, when people with AIDS die it is from other diseases such as tuberculosis (too bur kyuh LOH sus), pneumonia, or cancer.

From 1981 to 1999, more than 724,000 cases of AIDS were documented in the United States. At this time the disease has no known cure. However, several medications help treat AIDS in some patients. One group of medicines, such as AZT, interferes with the way that the virus multiplies in the host cell and is effective if it is used in the early stages of the disease. Another group of medicines that is being tested blocks the entrance of HIV into the host cell. These medicines prevent the pathogen from binding to the cell's surface.

SCIENCE *Online*

Data Update Visit the Glencoe Science Web site at **science.glencoe.com** for recent data about the the number of AIDS cases worldwide. Communicate to your class what you learn.

Figure 10
A person can be infected with HIV and not show any symptoms of the infection for several years. *Why does this characteristic make the spread of AIDS more likely?*

Magnification: 130,000×

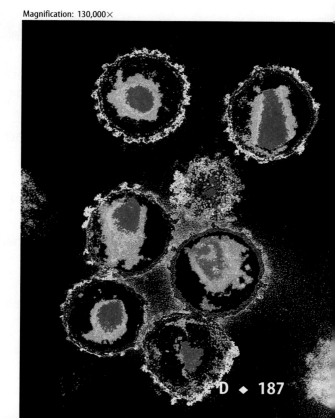

Fighting Disease

Washing a small wound with soap and water is the first step in preventing an infection. Cleaning the wound with an antiseptic and covering it with a bandage are other steps. Is it necessary to wash your body to help prevent diseases? Yes! In addition to reducing body odor, washing your body removes and destroys some surface microorganisms. In medical facilities, hand washing as shown in **Figure 11,** is important to reduce the spread of pathogens. It is also important for everyone to wash his or her hands to reduce the spread of disease.

In your mouth, microorganisms are responsible for mouth odor and tooth decay. Using dental floss and routine tooth brushing keep these organisms under control.

Figure 11
Proper hand washing includes using warm water and soap. The soapy lather must be rubbed over the hands, wrists, fingers, and thumbs for 15-20 s. Thoroughly rinse and dry with a clean towel.

Exercise and good nutrition help the circulatory and respiratory systems work more effectively. Good health habits, including getting enough rest and eating well-balanced meals, can make you less susceptible to the actions of disease organisms such as those that cause colds and flu. Keeping up with recommended immunizations and having annual health checkups also can help you stay healthy.

Section Assessment

1. How did the discoveries of Pasteur, Koch, and Lister help in the battle against the spread of disease?

2. List an infectious disease caused by each of the following: *a virus, a bacterium, a protist,* and *a fungus.*

3. How is the way HIV affects the immune system different from other viruses?

4. What are STDs? How are they contracted and treated?

5. **Think Critically** In what ways does Koch's procedure demonstrate the use of scientific methods?

Skill Builder Activities

6. **Recognizing Cause and Effect** How is poor cleanliness related to the spread of disease? Write your answer in your Science Journal. **For more help, refer to the** Science Skill Handbook.

7. **Making and Using Graphs** Make a bar graph using the following data about the number of deaths from AIDS-related diseases for children younger than 13 years old: *1995, 536; 1996, 420; 1997, 209; 1998, 115; and 1999, 76.* **For more help, refer to the** Science Skill Handbook.

Activity

Microorganisms and Disease

Microorganisms are everywhere. Washing your hands and disinfecting items you use helps remove some of these organisms.

What You'll Investigate

How do microorganisms cause infection?

Materials

fresh apples (6)
rotting apple
rubbing alcohol (5 mL)
self-sealing plastic bags (6)
labels and pencil
latex gloves

paper towels
sandpaper
cotton ball
soap and water
newspaper

Goals

■ **Observe** the transmission of microorganisms.
■ **Relate** microorganisms to infections.

Safety Precautions

WARNING: *Do not eat the apples. Do not remove goggles until the activity and cleanup are completed.* When you complete the experiment, give all bags to your teacher for disposal, then wash your hands.

Procedure

1. **Label** the plastic bags 1 through 6. Put on gloves. Place a fresh apple in bag 1.

2. Rub the rotting apple over the other five apples. This is your source of microorganisms. **WARNING:** *Don't touch your face.*

3. Put one apple in bag 2.

4. Hold one apple 1.5 m above the floor and drop it on a newspaper. Put it in bag 3.

5. Rub one apple with sandpaper. Place this apple in bag 4.

6. Wash one apple with soap and water. Dry it well. Put this apple in bag 5.

7. Use a cotton ball to spread alcohol over the last apple. Let it air dry. Place it in bag 6.

8. Seal all bags and put them in a dark place.

9. On day 3 and day 7, compare all of the apples without removing them from the bags. **Record** your observations in a data table.

Apple Observations		
Condition	Day 3	Day 7
1. Fresh		
2. Untreated		
3. Dropped		
4. Rubbed with sandpaper		
5. Washed with soap and water		
6. Covered with alcohol		

Conclude and Apply

1. How does this experiment relate to infections on your skin?

2. Why is it important to clean a wound?

*C*ommunicating Your Data

Prepare a poster illustrating the advantages of washing hands to avoid the spread of disease. Get permission to put the poster near a school rest room. **For more help, refer to the** Science Skill Handbook.

Noninfectious Diseases

As You Read

What You'll Learn

- **Define** noninfectious diseases and list causes of them.
- **Describe** the basic characteristics of cancer.
- **Explain** what happens during an allergic reaction.

Vocabulary
noninfectious disease
allergy
allergen
chemotherapy

Why It's Important
Knowing the causes of noninfectious diseases can help you understand their prevention and treatment.

Figure 12
Allergic reactions are caused by many things. A Hives are one kind of allergic reaction. B Some common substances stimulate allergic responses in people.

Chronic Disease

It's a beautiful, late-summer day. Flowers are blooming everywhere. You and your cousin hurry to get to the ballpark before the first pitch of the game. "Achoo!" Your cousin sneezes. Her eyes are watery and red. "Oh no! I sure don't want to catch that cold," you mutter. "I don't have a cold," she responds, "it's my allergies." Not all diseases are caused by pathogens. Diseases and disorders such as diabetes, allergies, asthma, cancer, and heart disease are **noninfectious diseases.** They are not spread from one person to another. Many are chronic (KRAH nihk). This means that they can last for a long time. Although some chronic diseases can be cured, others cannot.

Some infectious diseases can be chronic too. For example, deer ticks carry a bacterium that causes Lyme disease. This bacterium can affect the nervous system, heart, and joints for weeks to years. It can become chronic if not treated. Antibiotics will kill the bacteria, but some damage cannot be reversed.

Allergies

If you've had an itchy rash after eating a certain food, you probably have an allergy to that food. An **allergy** is an overly strong reaction of the immune system to a foreign substance. Many people have allergic reactions to cosmetics, shellfish, strawberries, peanuts, and insect stings. Most allergic reactions are minor, as shown in **Figure 12.** However, severe allergic reactions can occur, causing shock and even death if they aren't treated promptly.

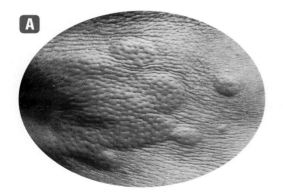

Allergens Substances that cause an allergic response are called **allergens.** Some chemicals, certain foods, pollen, molds, some antibiotics, and dust are allergens for some people. Some foods cause hives or stomach cramps and diarrhea. Pollen can cause a stuffy nose, breathing difficulties, watery eyes, and a tired feeling in some people. Dust can contain cat and dog dander and dust mites, as shown in **Figure 13.** Asthma (AZ muh) is a lung disorder that is associated with reactions to allergens. A person with asthma can have shortness of breath, wheezing, and coughing when he or she comes into contact with something they are allergic to.

When you come in contact with an allergen, your immune system usually forms antibodies. Your body reacts by releasing chemicals called histamines (HIHS tuh meenz) that promote red, swollen tissues. Antihistamines are medications that can be used to treat allergic reactions and asthma. Some severe allergies are treated with repeated injections of small doses of the allergen. This allows your body to become less sensitive to the allergen.

 **Reading Check** *What does your body release in response to an allergen?*

Diabetes

A chronic disease associated with the levels of insulin produced by the pancreas is diabetes. Insulin is a hormone that enables glucose to pass from the bloodstream into your cells. Doctors recognize two types of diabetes—Type 1 and Type 2. Type 1 diabetes is the result of too little or no insulin production. In Type 2 diabetes, your body cannot properly process the insulin. Symptoms of diabetes include fatigue, excessive thirst, frequent urination, and tingling sensations in the hands and feet.

If glucose levels in the blood remain high for a long time, health problems can develop. These problems can include blurred vision, kidney failure, heart attack, stroke, loss of feeling in the feet, and the loss of consciousness (diabetic coma). Patients with Type 1 diabetes, as shown in **Figure 14,** must monitor their intake of sugars and usually require daily injections of insulin to control their glucose levels. Careful monitoring of diet and weight usually are enough to control Type 2 diabetes. Since 1980, there has been an increase in the number of people with diabetes. Although the cause of diabetes is unknown, scientists have discovered that Type 2 diabetes is more common in people who are overweight and that it might be inherited.

Magnification: 245×

Figure 13
Dust mites are smaller than a period at the end of a sentence. They can live in pillows, mattresses, carpets, furniture, and other places.

Figure 14
Type 1 diabetes requires daily monitoring by either checking the amount of glucose in blood or the amount excreted in urine.

Chemicals and Disease

Chemistry INTEGRATION

Chemicals are everywhere—in your body, the foods you eat, cosmetics, cleaning products, pesticides, fertilizers, and building materials. Of the thousands of chemical substances used by consumers, less than two percent are harmful. Those chemicals that are harmful to living things are called toxins, as shown in **Figure 15.** Toxins can cause birth defects, cell mutations, cancers, tissue damage, chronic diseases, and death.

The Effects The amount of a chemical that is taken into your body and how long your body is in contact with it determine how it affects you. For example, low levels of a toxin might cause cardiac or respiratory problems. However, higher levels of the same toxin might cause death. Some chemicals, such as the asbestos shown in **Figure 15C,** can be inhaled over a long period of time. Eventually, the asbestos can cause chronic diseases of the lungs. Lead-based paints, if ingested, can accumulate in your body and eventually cause damage to the central nervous system. Another toxin, ethyl (EH thul) alcohol, is found in beer, wine, and liquor. It can cause birth defects in the children of mothers who drink alcohol during pregnancy.

Manufacturing, mining, transportation, and farming produce chemical wastes. These chemical substances interfere with the ability of soil, water, and air to support life. Pollution, caused by harmful chemicals, sometimes produces chronic diseases in humans. For example, long-term exposure to carbon monoxide, sulfur oxides, and nitrogen oxides in the air might cause a number of diseases, including bronchitis, emphysema (em fuh ZEE muh), and lung cancer.

Figure 15
Toxins can be in the environment.

A Chemical spills can be dangerous and might end up in groundwater.

B These scientists are testing the contents of barrels found in a dump.

C Asbestos, if inhaled into the lungs over a long period of time, can cause chronic diseases of the lungs. Protective clothing must be worn when removing asbestos.

Table 4 Characteristics of Cancer Cells

Cell growth is uncontrolled.
These cells do not function as part of your body.
The cells take up space and interfere with normal bodily functions.
The cells travel throughout your body.
The cells produce tumors and abnormal growths anywhere in your body.

Cancer

Cancer has been a disease of humans since ancient times. Egyptian mummies show evidence of bone cancer. Ancient Greek scientists described several different kinds of cancers. Even medieval manuscripts report details about the disease.

Cancer is the name given to a group of closely related diseases that result from uncontrolled cell growth. It is a complicated disease, and no one fully understands how cancers form. Characteristics of cancer cells are shown in **Table 4.** Certain regulatory molecules in the body control the beginning and ending of cell division. If this control is lost, a mass of cells called a tumor (TEW mur) results from this abnormal growth. Tumors can occur anywhere in your body. Cancerous cells can leave a tumor, spread throughout the body via blood and lymph vessels, and then invade other tissues.

> ✓ **Reading Check** *How do cancers spread?*

Types of Cancers Cancers can develop in any body tissue or organ. Leukemia (lew KEE mee uh) is a cancer of white blood cells. The cancerous white blood cells are immature and are no longer effective in fighting disease. The cancer cells multiply in the bone marrow and crowd out red blood cells, normal white blood cells, and platelets. Cancer of the lungs often starts in the bronchi and then spreads into the lungs. The surface area for air exchange in the lungs is reduced and breathing becomes difficult. Colorectal cancer, or cancer of the large intestine, is one of the leading causes of death among men and women. Changes in bowel movements and blood in the feces may be indications of the disease. In breast cancer, tumors grow in the breast. The second most common cancer in males is cancer of the prostate gland, which is an organ that surrounds the urethra.

Environmental Science

INTEGRATION

Dioxin is a dangerous chemical found in small amounts in certain herbicides. It can cause miscarriages, cancers, and liver disorders. Research to find out about the dioxin contamination in Times Beach, Missouri. Write a brief report in your Science Journal.

Figure 16
Tobacco products have been linked directly to lung cancer. Some chemicals around the home are carcinogenic.

Causes In the latter part of the eighteenth century, a British physician recognized the association of soot to cancer in chimney sweeps. Since that time, scientists have learned more about causes of cancer. Research done in the 1940s and 1950s related genes to cancer.

Although not all the causes of cancer are known, many causes have been identified. Smoking has been linked to lung cancer. Lung cancer is the leading cause of cancer deaths for males in the United States. Exposure to certain chemicals also can increase your chances of developing cancer. These substances, called carcinogens, (kar SIH nuh junz) include asbestos, various solvents, heavy metals, alcohol, and home and garden chemicals, as shown in **Figure 16.**

Exposure to X rays, nuclear radiation, and ultraviolet radiation of the Sun also increases your risk of getting cancer. Exposure to ultraviolet radiation might lead to skin cancer. Certain foods that are cured, or smoked, including barbecued meats, can give rise to cancers. Some food additives and certain viruses are suspected of causing cancers. Some people have a genetic predisposition for cancer, meaning that they have genes that make them more susceptible to the disease. This does not mean that they definitely will have cancer, but if it is triggered by certain factors they have a greater chance of developing cancer.

Treatment Surgery to remove cancerous tissue, radiation with X rays to kill cancer cells, and chemotherapy are some treatments for cancer. **Chemotherapy** (kee moh THUR uh pee) is the use of chemicals to destroy cancer cells. However, early detection of cancer is the key to any successful treatment.

Research in the science of immune processes, called immunology, has led to some new approaches for treating cancer. For example, specialized antibodies produced in the laboratory are being tested as anticancer agents. These antibodies are used as carriers to deliver medicines and radioactive substances directly to cancer cells. In another test, killer T cells are removed from a cancer patient and treated with chemicals that stimulate T cell production. The treated cells are then reinjected into the patient. Trial tests have shown some success in destroying certain types of cancer cells with this technique.

Prevention Knowing some causes of cancer might help you prevent it. The first step is to know the early warning signs, shown in **Table 5.** Medical attention and treatments such as chemotherapy or surgery in the early stages of some cancers can cure or keep them inactive.

A second step in cancer prevention concerns lifestyle choices. Choosing not to use tobacco and alcohol products can help prevent mouth and lung cancers and the other associated respiratory and circulatory system diseases. Selecting a healthy diet without many foods that are high in fats, salt, and sugar also might reduce your chances of developing cancer. Using sunscreen lotions and limiting the amount of time that you expose your skin to direct sunlight are good preventive measures against skin cancer. Careful handling of harmful home and garden chemicals will help you avoid the dangers connected with these substances. Carefully read the entire label before you use any product.

Inhaling certain air pollutants such as carbon monoxide, sulfur dioxide, nitric oxide, and asbestos fibers is dangerous to your health. To keep the air you breathe cleaner, the U.S. Government has regulations such as the Clean Air Act. These laws are intended to reduce the amount of these substances that are released into the air.

Table 5   **Early Warning Signs of Cancer** (from the National Cancer Institute)
Changes in bowel or bladder habits
A sore that does not heal
Unusual bleeding or discharge
Thickening or lump in the breast or elsewhere
Indigestion or difficulty swallowing
Obvious change in a wart or mole
Nagging cough or hoarseness

Section Assessment

1. Explain why diabetes is classified as a non-infectious disease.
2. Describe two ways cancer cells affect body organ functions.
3. Relate two causes of cancer to two methods to prevent cancer.
4. What are some ways your body can respond to allergens?
5. **Think Critically** Joel has an ear infection. The doctor prescribes an antibiotic. After taking the antibiotic, Joel breaks out in a rash. What is happening to him?

Skill Builder Activities

6. **Making and Using Tables** Make a table that relates several causes of cancer and their effects on your body. **For more help, refer to the** Science Skill Handbook.
7. **Using a Database** Use references to find information on different allergens. Group the allergens into the following categories: *chemical, food, mold, pollen,* and *antibiotic*. Use a computer to make a database. Which group has the most allergens? **For more help, refer to the** Technology Skill Handbook.

Activity
Design Your Own Experiment

Defensive Saliva

What happens when you think about a juicy cheeseburger or smell freshly baked bread? Your mouth starts making saliva. Saliva is the first line of defense for fighting harmful bacteria, acids, and bases entering your body. Saliva contains salts and chemicals known as bicarbonates. An example of a bicarbonate found in your kitchen is baking soda. Bicarbonates help to maintain normal pH levels in your mouth. When surfaces in your mouth have normal pH levels, the growth of bacteria is slowed and the effects of acids and bases are reduced. In this activity, you will design your own experiment to show the importance of saliva bicarbonates.

Recognize the Problem

How do the bicarbonates in saliva work to protect your mouth from harmful bacteria, acids, and bases?

Form a Hypothesis

Based on your reading in the text, form a hypothesis about how the bicabonates in saliva react to acids and bases.

Goals
- **Design** an experiment to test the reaction of a bicarbonate to acids and bases.
- **Test** the reaction of a bicarbonate to acids and bases.

Possible Materials
head of red cabbage   bicarbonate of soda
cooking pot   water
coffee filter   spoon
drinking glasses   white vinegar
clear household   lemon juice
   ammonia   orange juice

Safety Precautions

WARNING: *Never eat or drink anything used in an investigation.*

Test Your Hypothesis

Plan

1. **List** the materials you will need for your experiment. Red cabbage juice can be used as an indicator to test for acids and bases. Vinegar and citrus juices are acids, ammonia is a base, and baking soda (bicarbonate of soda) is a bicarbonate.

2. **Describe** how you will prepare the red cabbage juice and how you will use it to test for the presence of acids and bases.

3. **Describe** how you will test the effect of bicarbonate on acids and bases.

4. **List** the steps you will take to set up and complete your experiment. Describe exactly what you will do in each step.

5. Prepare a data table in your Science Journal to record your observations.

6. Examine the steps of your experiment to make certain they are in logical order.

Do

1. Ask your teacher to examine the steps of your experiment and data table before you start.

2. Conduct your experiment according to the approved plan.

3. **Record** your observations in your data table.

Analyze Your Data

1. **Compare** the color change of the acids and bases in the cabbage juice.

2. **Describe** how well the bicarbonate neutralized the acids and bases.

3. **Identify** any problems you had while setting up and conducting your experiment.

Draw Conclusions

1. Did your results support your hypothesis?

2. Based on your experiment, explain why your saliva contains a bicarbonate.

3. **Predict** how quickly bacteria would grow in your glass containing acid compared to another glass containing acid and the bicarbonate.

4. **Explain** how saliva protects your mouth from bacteria.

5. **Predict** what would happen if your saliva were made of only water.

*C*ommunicating
Your Data

Using what you learned in this experiment, create a poster about the importance of good dental hygiene. Invite a dental hygienist to speak to your class.

Science Stats

Battling Bacteria

Did you know...

...**The term** *antibiotic* was first coined by an American microbiologist. The scientist received a Nobel prize in 1952 for the discovery of streptomycin (strep toh MY suhn), an antibiotic used against tuberculosis.

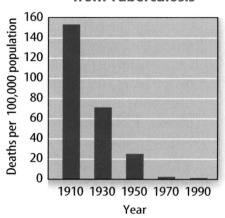

United States Death Rates from Tuberculosis

(Bar graph: y-axis "Deaths per 100,000 population" ranging 0 to 160; x-axis "Year" with values 1910, 1930, 1950, 1970, 1990. Bar heights: 1910 ≈ 155, 1930 ≈ 70, 1950 ≈ 25, 1970 ≈ 2, 1990 ≈ 1.)

State of North Dakota
PRESCRIPTION BLANK

Mary Jackson, M.D.
General Medicine
265 Main Street
Greenville, ND

Patient *Julia Ramirn*   DOB *12/04/65*
Addr. *62 North St. Greenville*   Date *11/06/01*

℞   *Amoxicillin 250mg tabs*
    Sig: † tab po q8h

Substitution permissible ✓   Do Not ____
Do Not Refill ✓   Substitute
Refill ____   Signature *Mary Jackson*
Times ____

...**One of the frequently prescribed drugs** is the antibiotic amoxicillin (uh mahk see SI luhn), a chemical variation of penicillin. It is prescribed for a variety of infections. In 1998 alone, amoxicillin was prescribed 16.7 million times, comprising 1.4 percent of the total prescriptions written that year.

...**Not all bacteria are harmful.** Our bodies contain millions of helpful bacteria that promote digestion, produce B vitamins, and crowd out bacteria that cause disease. Bacterial cells outnumber human cells in your body.

...**People have long used natural remedies** to treat infections. These remedies include garlic, *Echinacea* (purple coneflower), and an antibiotic called squalamine, found in sharks' stomachs.

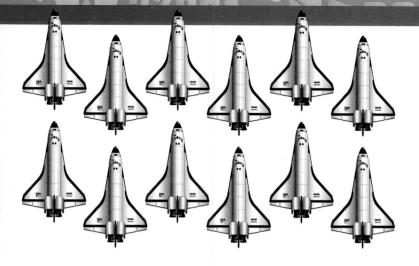

...Pharmaceutical companies in the United States produce nearly 23 million kg of antibiotics each year. That's equivalent to the weight of about 50 space shuttles. In 1954, these companies produced only about 90,000 kg of antibiotics.

Antibiotics Prescribed Each Year in the United States

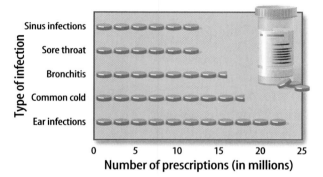

...In recent decades many bacteria have become resistant to antibiotics. For example, one group of bacteria that cause illnesses of the stomach and intestines—*Shigella* (shih GEL uh)—became harder to control. In 1985, less than one third of *Shigella* were resistant to the antibiotic ampicillin (am puh SI luhn). By 1991, however, more than two thirds of *Shigella* could continue to grow in the presence of the drug.

Do the Math

1. Calculate the percentage by which antibiotic production in the United States has increased from 1954 to today.
2. An estimated 100 million doses of antibiotics are prescribed each year. With 273 million people living in the United States, what is the average number of doses per person?
3. It is believed that 30 percent of the antibiotics prescribed for ear infections are unnecessary. Using the graph, calculate the number of unnecessary prescriptions.

Go Further

Go to **science.glencoe.com** to research the production of four antibiotics. Create a graph comparing the number of kilograms of each antibiotic produced in one year.

Reviewing Main Ideas

Section 1 The Immune System

1. Your body is protected against most pathogens by the immune system, which includes skin, cilia and mucus in the respiratory system, white blood cells in the circulatory system, and enzymes and hydrochloric acid in the digestive system. The purpose of the immune system is to fight disease.

2. Active immunity is long lasting, but passive immunity is not. *What are other ways to prevent the spread of disease?*

3. Antigens are complex molecules that identify foreign molecules in your body. Your body makes an antibody that attaches to a specific antigen, making it harmless.

Section 2 Infectious Diseases

1. Pasteur and Koch discovered that microorganisms cause diseases. Lister learned that cleanliness helps control microorganisms.

2. Air, water, food, and animal contact can pass a pathogen from one person to another. Bacteria, viruses, fungi, and protists can cause infectious diseases. *How could this mosquito pass on disease?*

3. Sexually transmitted diseases (STDs) can be passed between persons during sexual contact. They include genital herpes, gonorrhea, syphilis, and AIDS.

4. HIV can be transmitted by sexual contact, by using a disease-contaminated needle, by transfusion with contaminated blood, and to a fetus from its mother. AIDS damages your body's immune system so that it cannot fight infections.

Section 3 Noninfectious Diseases

1. Causes of noninfectious diseases include genetics, chemicals, poor diet, and uncontrolled cell growth. Chronic noninfectious diseases include diabetes, cancer, heart disease, and allergies.

2. An allergy is a reaction of the immune system to a foreign substance. Your body releases histamines that cause red, swollen tissues. *What are substances that can cause rashes called?*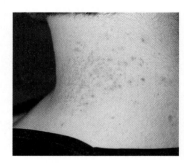

3. Cancer results from uncontrolled cell growth. When this control is lost, the cells multiply, spread through the blood and lymph vessels, and invade normal tissues.

4. Some cancers can be cured or kept in remission by medical intervention if they are detected early. Cancer is treated with surgery, chemotherapy, and radiation. Lifestyle choices can help prevent cancer.

FOLDABLES
Reading & Study Skills

After You Read

Based on information in this chapter, circle the diseases on your Foldable that are the greatest threat to people on Earth.

Visualizing Main Ideas

Complete the following concept map on infectious diseases.

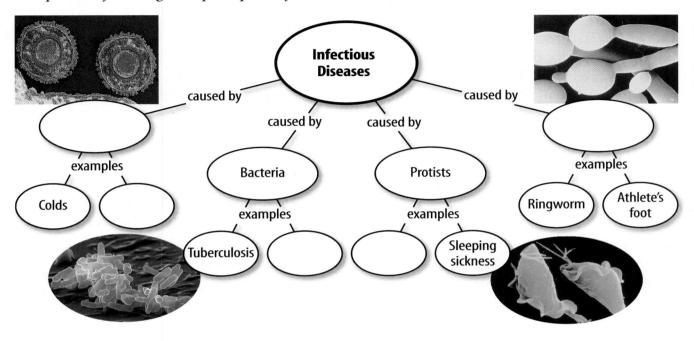

caused by

Infectious Diseases

caused by   caused by   caused by

examples   Bacteria   Protists   examples

Colds   Ringworm   Athlete's foot

examples   examples

Tuberculosis   Sleeping sickness

Vocabulary Review

Vocabulary Words

a. active immunity
b. allergen
c. allergy
d. antibody
e. antigen
f. biological vector
g. chemotherapy
h. immune system
i. infectious disease
j. noninfectious disease
k. passive immunity
l. pasteurization
m. sexually transmitted disease (STD)
n. vaccination
o. virus

Study Tip

Keep all your homework assignments, and read them from time to time. Make sure you understand any questions that you may have answered incorrectly.

Using Vocabulary

Replace each underlined word with the correct vocabulary word.

1. An <u>allergen</u> can cause infectious diseases.

2. A disease-carrying organism is called a <u>noninfectious disease</u>.

3. Measles is an example of <u>pastueurization</u>.

4. Injection of weakened viruses is called <u>biological vector</u>.

5. <u>Passive immunity</u> occurs when your body makes its own antibodies.

6. An <u>antigen</u> stimulates histamine release.

7. Heating a liquid to kill harmful bacteria is called <u>chemotherapy</u>.

8. Diabetes is an example of a <u>sexually transmitted</u> disease.

Checking Concepts

Choose the word or phrase that best answers the question.

1. How do scientists know that a pathogen causes a specific disease?
   A) It is present in all cases of the disease.
   B) It does not infect other animals.
   C) It causes other diseases.
   D) It is treated with heat.

2. How can infectious diseases be caused?
   A) heredity   C) chemicals
   B) allergies   D) organisms

3. Which of the following might be a biological vector?
   A) bird   C) water
   B) rock   D) soil

4. What is formed in the blood to fight invading antigens?
   A) hormones   C) pathogens
   B) allergens   D) antibodies

5. Which of the following is one of your body's general defenses against some pathogens?
   A) stomach enzymes   C) some vaccines
   B) HIV   D) hormones

6. Which of the following is known as an infectious disease?
   A) allergies   C) syphilis
   B) asthma   D) diabetes

7. Which disease is caused by a virus that attacks white blood cells?
   A) AIDS   C) flu
   B) measles   D) polio

8. Which of the following is a characteristic of cancer cells?
   A) controlled cell growth
   B) help your body stay healthy
   C) interfere with normal body functions
   D) do not multiply or spread

9. Which of the following is caused by a virus?
   A) AIDS   C) ringworm
   B) gonorrhea   D) syphilis

10. How can cancer cells be destroyed?
    A) chemotherapy   C) vaccines
    B) antigens   D) viruses

Thinking Critically

11. Is it better to vaccinate people or to wait until they build their own immunity? Explain.

12. What advantage might a breast-fed baby have compared to a bottle-fed baby?

13. How does your body protect itself from antigens?

14. How do helper T cells and B cells work to eliminate antigens?

15. Describe the differences among antibodies, antigens, and antibiotics.

Developing Skills

16. **Making and Using Tables** Complete this table.

Comparing Diseases		
Disease	Cause	Prevention
Cancer		
Tetanus		
Measles		

17. **Recognizing Cause and Effect** Use library references to identify the cause—bacteria, virus, fungus, or protist—of each of these diseases: athlete's foot, AIDS, cold, dysentery, flu, pinkeye, acne, and strep throat.

18. **Classifying** Using word processing software, make a table to classify the following diseases as infectious or noninfectious: diabetes, gonorrhea, herpes, strep throat, syphilis, cancer, and flu.

19. **Interpreting Data** Using the graph below, explain the rate of polio cases between 1952 and 1965. What conclusions can you draw about the effectiveness of the polio vaccines?

Cases of Polio

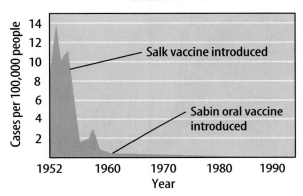

20. **Concept Mapping** Make a network tree concept map that compares the various defenses your body has against diseases. Compare general defenses, active immunity, and passive immunity.

Performance Assessment

21. **Poster** Design and construct a poster to illustrate how a person with the flu could spread the disease to family members, classmates, and others.

TECHNOLOGY

Go to the Glencoe Science Web site at **science.glencoe.com** or use the **Glencoe Science CD-ROM** for additional chapter assessment.

 **Test Practice**

Mrs. Henson showed her class a graph about life expectancy between the years 1970 and 1997.

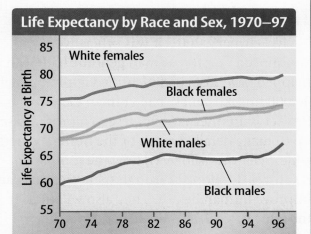

Study the graph and answer the following questions.

1. According to the information in the graph, which group had the highest life expectancy in both 1975 and 1994?
   A) white males   **C)** white females
   B) black females   **D)** black males

2. A reasonable hypothesis based on the information in the graph is that _____ .
   F) life expectancy has decreased in the period between 1970 and 1994
   G) life expectancy of white females is the lowest because they suffer the most disease
   H) females usually live shorter lives than males
   J) life expectancy has slowly increased in the period between 1970 and 1994

Reading Comprehension

Read the passage carefully. Then read the questions that follow the passage. Decide which is the best answer to each question.

The Human Genome Project

In February, 2001, it was officially announced that the human genome had been mapped. Scientists had identified approximately 30,000 human genes. The Human Genome Project is an international research program set up to make detailed maps of the human genetic code. Scientists will learn how the tools and resources developed through their research can improve human health. They also will look at the ethical, legal, and social concerns their research brings up.

The human genome is comprised of all the DNA that exists in human beings. This includes all of your genes. The DNA in your genes carries information on how to manufacture the proteins that are required for human life. These proteins determine many of your characteristics, such as how you look, how well you fight off infections, how you digest food, and the color of your hair. Your genes are located on 23 pairs of chromosomes in the nucleus of each of your cells. Many diseases such as cystic fibrosis, sickle-cell anemia, and Parkinson's disease are passed on from generation to generation through the information in genes.

The completion of the Human Genome Project might help scientists develop medicines that could cure hereditary diseases. The information gathered by the Human Genome Project also might help doctors identify and alert patients who are at risk of certain <u>inherited</u> diseases. With knowledge and awareness, doctors and patients might be able to take steps to prevent such diseases from developing.

Scientists hope that the Human Genome Project will advance medical science into a new era of diagnosing, preventing, treating, and curing diseases.

Test-Taking Tip   When you read a passage, make a list of vocabulary words that you do not understand.

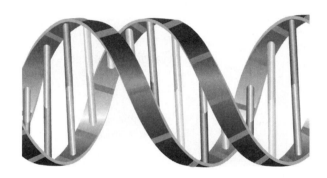

Watson and Crick's double-helix model of DNA

1. The term <u>inherited</u> in this passage best means _____.
   A) averted
   B) acquired
   C) inborn
   D) lost

2. One of the main ideas of the passage is that _____.
   F) human genes are located on the 23 pairs of chromosomes
   G) the Human Genome Project might help doctors diagnose, prevent, and cure diseases
   H) the Human Genome Project is a historic scientific achievement
   J) genes are responsible for many diseases, such as cystic fibrosis, sickle-cell anemia, and Parkinson's disease

Reasoning and Skills

Read each question and choose the best answer.

NEURONS   DENDRITES   AXON

1. Which of the following belongs with the group above?
A) arteries
B) synapses
C) tendons
D) lymph nodes

Test-Taking Tip   Think about which human body system the words in the box relate to.

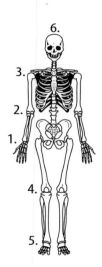

6.
3.
2.
1.
4.
5.

2. The place where two or more bones come together is called a joint. Joints can be movable or immovable. Which joint is an example of an immovable joint?
F) 2   H) 5
G) 3   J) 6

Test-Taking Tip   Think about which joints move when you move.

3. Laticia has just jogged for two miles. When she stops, her heart rate and her breathing rate are higher than normal and have increased to 150 beats per minute and 40 breaths per minute. The reason why her heart rate and her breathing rate are high is because _____.
A) her body is fighting off an infection
B) her muscles and cells need more oxygen and nutrients
C) her nervous system is reacting to a lot of stimuli
D) she is not in good physical shape

Test-Taking Tip   Consider the responsibilities of the human circulatory and respiratory systems.

Consider this question carefully before writing your answer on a separate sheet of paper.

4. Doctors recommend that you eat a well-balanced diet that contains food from each category of the food pyramid and that you consume an appropriate number of Calories per day. What are some reasons why it is important for you to eat a well-balanced diet?

Test-Taking Tip   Think about the effects that eating healthy and unhealthy food has on the human body.

Student Resources

Student Resources

CONTENTS

Field Guides 208

Emergencies Field Guide 208

Skill Handbooks 212

Science Skill Handbook 212

Organizing Information 212
Researching Information 212
Evaluating Print and
Nonprint Sources 212
Interpreting Scientific Illustrations . . 213
Venn Diagram 213
Concept Mapping 213
Writing a Paper 215

Investigating and Experimenting 216
Identifying a Question 216
Forming Hypotheses. 216
Predicting. 216
Testing a Hypothesis 216
Identifying and Manipulating
Variables and Controls 217
Collecting Data 218
Measuring in SI 219
Making and Using Tables 220
Recording Data 221
Recording Observations 221
Making Models 221
Making and Using Graphs 222

Analyzing and Applying Results 223
Analyzing Results. 223
Forming Operational Definitions . . . 223
Classifying 223
Comparing and Contrasting 224
Recognizing Cause and Effect 224
Interpreting Data. 224
Drawing Conclusions 224

Evaluating Others' Data and
Conclusions 225
Communicating 225

Technology Skill Handbook 226
Using a Word Processor 226
Using a Database 226
Using an Electronic Spreadsheet 227
Using a Computerized
Card Catalog 228
Using Graphics Software 228
Developing Multimedia
Presentations 229

Math Skill Handbook 230
Converting Units. 230
Using Fractions 231
Calculating Ratios 232
Using Decimals 232
Using Percentages 233
Using Precision and
Significant Digits. 233
Solving One-Step Equations 234
Using Proportions. 235
Using Statistics. 236

Reference Handbook 237

A. Care and Use of a Microscope. 237
B. Safety in the Science Classroom 238
C. SI/Metric to English,
English to Metric Conversions 239
D. Diversity of Life. 240

English Glossary. 244
Spanish Glossary 250
Index 257

Field GUIDE

Emergencies

Use your senses and look for these signs.

W hen you hear a scream or see smoke, do you wonder if you are witnessing an emergency? Emergencies can be difficult to identify. The scream you hear could be a hungry baby, and the smoke could be coming from a neighbor's grill, but sometimes the emergency is real. You might be the only one around to help. Knowing ahead of time how to identify an emergency will help you decide what to do when you are in an emergency situation.

Recognizing an Emergency

Emergencies can happen any-time and anyplace. A teammate might slip and twist her ankle dur-ing soccer practice. How can you tell whether the situation is a true emergency?

It can be hard to know whether the situation is an emergency. If you decide that it is, then you must determine the best way to help. Taking action is the first step to helping in an emergency. This can be difficult because you might be afraid of becoming involved. Calling the local emergency telephone number for help is important. If there isn't a tele-phone nearby, you can help by getting an adult.

Noises

You know there could be an emergency when you hear screaming, calls for help, breaking glass, sudden or loud noises made by things or people, or screeching tires.

Sights

If you see a stalled vehicle or a vehicle off the road, smoke or fire, or a person lying uncon-scious on the floor or on the ground, you know there's an emergency.

Field Activity

Visit the Glencoe Science Web site at **science.glencoe.com** to connect to the National Red Cross site where you can get further information on emergencies. Record this information in your Science Journal. Set up posters at school showing how to recognize each of them.

Odors

If you detect different or strange odors or familiar odors that are stronger than usual, it might be a signal that there's a problem. Leave the area immediately if you notice these unusual or very strong odors. They might be poisonous and extremely dangerous.

Behavior or Appearance of Another Person

If someone you know or someone you see shows any of the following signs, treat the situation as an emergency: unconsciousness, severe bleeding, vomiting or passing blood, difficulty breathing, grabbing his or her own throat, unexplained confusion or drowsiness, sudden change in skin color, complaining of chest pains or pressure, seizure, severe headache, or slurred speech.

Signs and symptoms vary from person to person, so you still might be unsure if it's a true emergency. In this case, pay attention to your instincts. Do you feel that getting involved might be dangerous for you? If so, stay away and try to get help. Do you feel that someone's safety depends on you and that you can help safely? If so, it's time to respond. **WARNING:** *If you become injured, you can't help.*

Field GUIDE

Responding to an Emergency

After you decide that the situation is an emergency, you should follow three basic steps: **check, call,** and **care.**

CHECK

When you call for help, you will need to be able to pass on as much information as possible about the situation to the emergency operator. Your first priority is to identify the people who are hurt or in trouble. Sometimes, a quiet victim can be overlooked if another is screaming. Get help from others. Bystanders might tell you exactly what happened or the location of the nearest phone. They also might help comfort victims or apply first aid if they are qualified. Be conscious of signs that tell you what might have caused the emergency, such as broken glass, heavy objects, or odors.

CALL

The most important first-aid tool at your disposal is the telephone. In most but not all areas, the emergency number is 911. Research your local emergency number and memorize it. **Never hesitate to call for help when you are in an emergency.**

When speaking with an emergency operator, remain calm and answer each question as clearly as possible. Stay on the line until the operator tells you to hang up.

CARE

If you are trained in first aid, follow standard procedures. Otherwise, do not perform any procedures on victims. Your job is to stay with the victims until help arrives. Here are some pointers. Respond to victims with life-threatening injuries or illnesses first. **WARNING:** *Do not move victims unless they are in danger* from an explosion, fire, or poisonous gas. Before helping a victim, get his or her permission and explain what you are going to do. If the person declines your help, do not insist. Keep victims comfortable, talk to them, and reassure them. Prevent victims from becoming chilled or overheated.

If you remember the 3 Cs—**check, call,** and **care**—someone who is injured or ill might have a better chance of surviving or recovering more quickly. What can you do right now to prepare for the emergency that might occur tomorrow, next week, or next year?

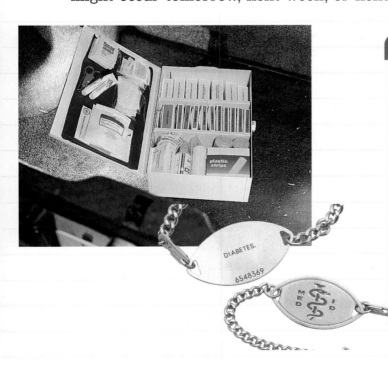

Emergency Checklist

Place first-aid kits in your home, car, and garage. Post emergency numbers near all phones and tape them to the first-aid kits. Teach children how to call for help. Find out whether your town has a 911 system. Put easy-to-read street numbers on your home or apartment. If you have special medical needs, wear a medical alert tag. Learn first-aid techniques including when and how to use CPR (cardio-pulmonary resuscitation) and the abdominal thrust for choking victims.

Organizing Information

As you study science, you will make many observations and conduct investigations and experiments. You will also research information that is available from many sources. These activities will involve organizing and recording data. The quality of the data you collect and the way you organize it will determine how well others can understand and use it. In **Figure 1,** the student is obtaining and recording information using a microscope.

Putting your observations in writing is an important way of communicating to others the information you have found and the results of your investigations and experiments.

Researching Information

Scientists work to build on and add to human knowledge of the world. Before moving in a new direction, it is important to gather the information that already is known about a subject. You will look for such information in various reference sources. Follow these steps to research information on a scientific subject:

Step 1 Determine exactly what you need to know about the subject. For instance, you might want to find out what happened to local plant life when Mount St. Helens erupted in 1980.

Step 2 Make a list of questions, such as: When did the eruption begin? How long did it last? How large was the area in which plant life was affected?

Step 3 Use multiple sources such as textbooks, encyclopedias, government documents, professional journals, science magazines, and the Internet.

Step 4 List where you found the sources. Make sure the sources you use are reliable and the most current available.

Figure 1
Making an observation is one way to gather information directly.

Evaluating Print and Nonprint Sources

Not all sources of information are reliable. Evaluate the sources you use for information, and use only those you know to be dependable. For example, suppose you want information about the digestion of fats and proteins. You might find two Websites on digestion. One Web site contains "Fat Zapping Tips" written by a company that sells expensive, high-protein supplements to help your body eliminate excess fat. The other is a Web page on "Digestion and Metabolism" written by a well-respected medical school. You would choose the second Web site as the more reliable source of information.

In science, information can change rapidly. Always consult the most current sources. A 1985 source about the human genome would not reflect the most recent research and findings.

Interpreting Scientific Illustrations

As you research a science topic, you will see drawings, diagrams, and photographs. Illustrations help you understand what you read. Some illustrations are included to help you understand an idea that you can't see easily by yourself. For instance, you can't see the bones of a blue whale, but you can look at a diagram of a whale skeleton as labeled in **Figure 2** that helps you understand them. Visualizing a drawing helps many people remember details more easily. Illustrations also provide examples that clarify difficult concepts or give additional information about the topic you are studying.

Most illustrations have a label or a caption. A label or caption identifies the illustration or provides additional information to better explain it. Can you find the caption or labels in **Figure 2?**

Figure 2
A labeled diagram of the skeletal structure of a blue whale

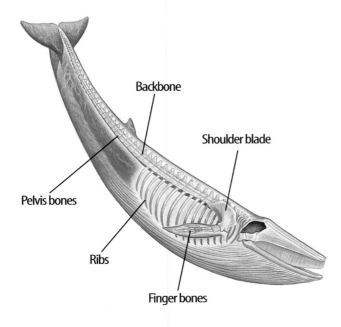

Backbone

Shoulder blade

Pelvis bones

Ribs

Finger bones

Venn Diagram

Although it is not a concept map, a Venn diagram illustrates how two subjects compare and contrast. In other words, you can see the characteristics that the subjects have in common and those that they do not.

The Venn diagram in **Figure 3** shows the relationship between two categories of organisms, plants and animals. Both share some basic characteristics as living organisms. However, there are differences in the ways they carry out various life processes, such as obtaining nourishment, that distinguish one from the other.

Concept Mapping

If you were taking a car trip, you might take some sort of road map. By using a map, you begin to learn where you are in relation to other places on the map.

A concept map is similar to a road map, but a concept map shows relationships among ideas (or concepts) rather than places. It is a diagram that visually shows how concepts are related. Because a concept map shows relationships among ideas, it can make the meanings of ideas and terms clear and help you understand what you are studying.

Overall, concept maps are useful for breaking large concepts down into smaller parts, making learning easier.

Figure 3
A Venn diagram shows how objects or concepts are alike and how they are different.

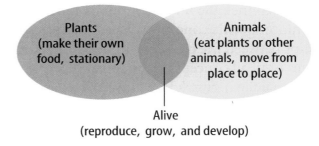

Plants
(make their own food, stationary)

Animals
(eat plants or other animals, move from place to place)

Alive
(reproduce, grow, and develop)

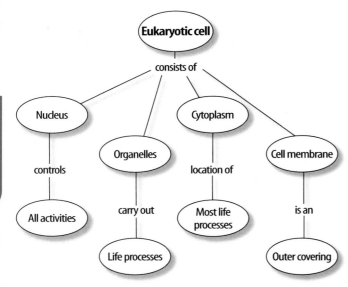

Figure 4
A network tree shows how concepts or objects are related.

Network Tree Look at the network tree in **Figure 4,** that shows details about a eukaryotic cell. A network tree is a type of concept map. Notice how some words are in ovals while others are written across connecting lines. The words inside the ovals are science terms or concepts. The words written on the connecting lines describe the relationships between the concepts.

When constructing a network tree, write the topic on a note card or piece of paper. Write the major concepts related to that topic on separate note cards or pieces of paper. Then arrange them in order from general to specific. Branch the related concepts from the major concept and describe the relationships on the connecting lines. Continue branching to more specific concepts. Write the relationships between the concepts on the connecting lines until all concepts are mapped. Then examine the network tree for relationships that cross branches, and add them to the network tree.

Events Chain An events chain is another type of concept map. It models the order of items or their sequence. In science, an events chain can be used to describe a sequence of events, the steps in a procedure, or the stages of a process.

When making an events chain, first find the one event that starts the chain. This event is called the *initiating event.* Then, find the next event in the chain and continue until you reach an outcome. Suppose you are asked to describe the main stages in the growth of a plant from a seed. You might draw an events chain such as the one in **Figure 5.** Notice that connecting words are not necessary in an events chain.

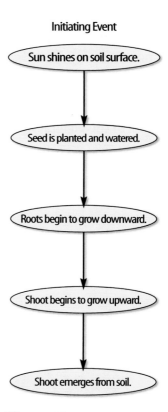

Figure 5
Events chains show the order of steps in a process or event.

Cycle Map A cycle concept map is a specific type of events chain map. In a cycle concept map, the series of events does not produce a final outcome. Instead, the last event in the chain relates back to the beginning event.

You first decide what event will be used as the beginning event. Once that is decided, you list events in order that occur after it. Words are written between events that describe what happens from one event to the next. The last event in a cycle concept map relates back to the beginning event. The number of events in a cycle concept varies but is usually three or more. Look at the cycle map in **Figure 6.**

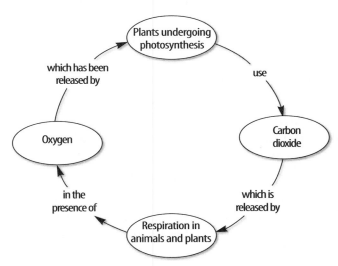

Figure 6
A cycle map shows events that occur in a cycle.

Spider Map A type of concept map that you can use for brainstorming is the spider map. When you have a central idea, you might find you have a jumble of ideas that relate to it but might not clearly relate to each other. The circulatory system spider map in **Figure 7** shows that if you write these ideas outside the main concept, then you can begin to separate and group unrelated terms so they become more useful.

Figure 7
A spider map allows you to list ideas that relate to a central topic but not necessarily to one another.

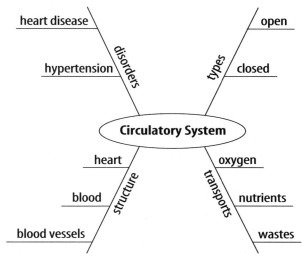

Writing a Paper

You will write papers often when researching science topics or reporting the results of investigations or experiments. Scientists frequently write papers to share their data and conclusions with other scientists and the public. When writing a paper, use these steps.

Step 1 Assemble your data by using graphs, tables, or a concept map. Create an outline.

Step 2 Start with an introduction that contains a clear statement of purpose and what you intend to discuss or prove.

Step 3 Organize the body into paragraphs. Each paragraph should start with a topic sentence, and the remaining sentences in that paragraph should support your point.

Step 4 Position data to help support your points.

Step 5 Summarize the main points and finish with a conclusion statement.

Step 6 Use tables, graphs, charts, and illustrations whenever possible.

You might say the work of a scientist is to solve problems. When you decide to find out why one corner of your yard is always soggy, you are problem solving, too. You might observe that the corner is lower than the surrounding area and has less vegetation growing in it. You might decide to see if planting some grass will keep the corner drier.

Scientists use orderly approaches to solve problems. The methods scientists use include identifying a question, making observations, forming a hypothesis, testing a hypothesis, analyzing results, and drawing conclusions.

Scientific investigations involve careful observation under controlled conditions. Such observation of an object or a process can suggest new and interesting questions about it. These questions sometimes lead to the formation of a hypothesis. Scientific investigations are designed to test a hypothesis.

Identifying a Question

The first step in a scientific investigation or experiment is to identify a question to be answered or a problem to be solved. You might be interested in knowing why an animal like the one in **Figure 8** looks the way it does.

Figure 8
When you see a bird, you might ask yourself, "How does the shape of this bird's beak help it feed?"

Forming Hypotheses

Hypotheses are based on observations that have been made. A hypothesis is a possible explanation based on previous knowledge and observations.

Perhaps a scientist has observed that bean plants grow larger if they are fertilized than if they are not. Based on these observations, the scientist can make a statement that he or she can test. The statement is a hypothesis. The hypothesis could be: *Fertilizer makes bean plants grow larger.* A hypothesis has to be something you can test by using an investigation. A testable hypothesis is a valid hypothesis.

Predicting

When you apply a hypothesis to a specific situation, you predict something about that situation. First, you must identify which hypothesis fits the situation you are considering. People use predictions to make everyday decisions. Based on previous observations and experiences, you might form a prediction that if fertilizer makes bean plants grow larger, then fertilized plants will yield more beans than plants not fertilized. Someone could use this prediction to plan to grow fewer plants.

Testing a Hypothesis

To test a hypothesis, you need a procedure. A procedure is the plan you follow in your experiment. A procedure tells you what materials to use, as well as how and in what order to use them. When you follow a procedure, data are generated that support or do not support the original hypothesis statement.

For example, suppose you notice that your guppies don't seem as active as usual when your aquarium heater is not working. You wonder how water temperature affects guppy activity level. You decide to test the hypothesis, "If water temperature increases, then guppy activity should increase." Then you write the procedure shown in **Figure 9** for your experiment and generate the data presented in the table below.

Procedure

1. Fill five identical glass containers with equal amounts of aquarium water.
2. Measure and record the temperature of the water in the first container.
3. Heat and cool the other containers so that two have higher and two have lower water temperatures.
4. Place a guppy in each container; count and record the number of movements each guppy makes in 5 minutes.

Figure 9
A procedure tells you what to do step by step.

Number of Guppy Movements		
Container	Temperature (°C)	Movements
1	38	56
2	40	61
3	42	70
4	36	46
5	34	42

Are all investigations alike? Keep in mind as you perform investigations in science that a hypothesis can be tested in many ways. Not every investigation makes use of all the ways that are described on these pages, and not all hypotheses are tested by investigations. Scientists encounter many variations in the methods that are used when they perform experiments. The skills in this handbook are here for you to use and practice.

Identifying and Manipulating Variables and Controls

In any experiment, it is important to keep everything the same except for the item you are testing. The one factor you change is called the independent variable. The factor that changes as a result of the independent variable is called the dependent variable. Always make sure you have only one independent variable. If you allow more than one, you will not know what causes the changes you observe in the dependent variable. Many experiments also have controls—individual instances or experimental subjects for which the independent variable is not changed. You can then compare the test results to the control results.

For example, in the guppy experiment, you made everything the same except the temperature of the water. The glass containers were identical. The volume of aquarium water in each container and beginning water temperature were the same. Each guppy was like the others, as much as possible. In this way, you could be sure that any difference in the number of guppy movements was caused by the temperature change—the independent variable. The activity level of the guppy was measured as the number of guppy movements—the dependent variable. The guppy in the container in which the water temperature was not changed was the control.

Collecting Data

Whether you are carrying out an investigation or a short observational experiment, you will collect data, or information. Scientists collect data accurately as numbers and descriptions and organize it in specific ways.

Observing Scientists observe items and events, then record what they see. When they use only words to describe an observation, it is called qualitative data. For example, a scientist might describe the color of a bird or the shape of a bird's beak as seen through binoculars. Scientists' observations also can describe how much there is of something. These observations use numbers, as well as words, in the description and are called quantitative data. For example, if a particular dog is described as being "furry, yellow, and short-haired," the data are clearly qualitative. Quantitative data for this dog might include "a mass of 14 kg, a height of 46 cm, and an age of 150 days." Quantitative data often are organized into tables. Then, from information in the table, a graph can be drawn. Graphs can reveal relationships that exist in experimental data.

When you make observations in science, you should examine the entire object or situation first, then look carefully for details. If you're looking at a plant, for instance, check general characteristics such as size and overall structure before using a hand lens to examine the leaves and other smaller structures such as flowers or fruits. Remember to record accurately everything you see.

Scientists try to make careful and accurate observations. When possible, they use instruments such as microscopes, metric rulers, graduated cylinders, thermometers, and balances. Measurements provide numerical data that can be repeated and checked.

Sampling When working with large numbers of objects or a large population, scientists usually cannot observe or study every one of them. Instead, they use a sample or a portion of the total number. To *sample* is to take a small, representative portion of the objects or organisms of a population for research. By making careful observations or manipulating variables within a portion of a group, information is discovered and conclusions are drawn that might apply to the whole population.

Estimating Scientific work also involves estimating. To *estimate* is to make a judgment about the size or the number of something without measuring or counting every object or member of a population. Scientists first count the number of objects in a small sample. Looking through a microscope lens, for example, a scientist can count the number of bacterial colonies in the 1-cm² frame shown in **Figure 10.** Then the scientist can multiply that number by the number of cm² in the petri dish to get an estimate of the total number of bacterial colonies present.

Figure 10
To estimate the total number of bacterial colonies that are present on a petri dish, count the number of bacterial colonies within a 1-cm² frame and multiply that number by the number of frames on the dish.

Measuring in SI

The metric system of measurement was developed in 1795. A modern form of the metric system, called the International System, or SI, was adopted in 1960. SI provides standard measurements that all scientists around the world can understand.

The metric system is convenient because unit sizes vary by multiples of 10. When changing from smaller units to larger units, divide by a multiple of 10. When changing from larger units to smaller, multiply by a multiple of 10. To convert millimeters to centimeters, divide the millimeters by 10. To convert 30 mm to centimeters, divide 30 by 10 (30 mm equal 3 cm).

Prefixes are used to name units. Look at the table below for some common metric prefixes and their meanings. Do you see how the prefix *kilo-* attached to the unit *gram* is *kilogram*, or 1,000 g?

Metric Prefixes			
Prefix	Symbol	Meaning	
kilo-	k	1,000	thousand
hecto-	h	100	hundred
deka-	da	10	ten
deci-	d	0.1	tenth
centi-	c	0.01	hundredth
milli-	m	0.001	thousandth

Now look at the metric ruler shown in **Figure 11.** The centimeter lines are the long, numbered lines, and the shorter lines are millimeter lines.

When using a metric ruler, line up the 0-cm mark with the end of the object being measured, and read the number of the unit where the object ends. In this instance it would be 4.5 cm.

Figure 11
This metric ruler shows centimeter and millimeter divisions.

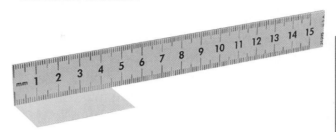

Liquid Volume In some science activities, you will measure liquids. The unit that is used to measure liquids is the liter. A liter has the volume of 1,000 cm³. The prefix *milli-* means "thousandth (0.001)." A milliliter is one thousandth of 1 L and 1 L has the volume of 1,000 mL. One milliliter of liquid completely fills a cube measuring 1 cm on each side. Therefore, 1 mL equals 1 cm³.

You will use beakers and graduated cylinders to measure liquid volume. A graduated cylinder, as illustrated in **Figure 12,** is marked from bottom to top in milliliters. This graduated cylinder contains 79 mL of a liquid.

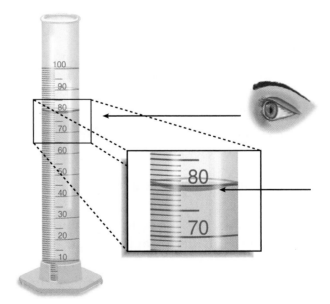

Figure 12
Graduated cylinders measure liquid volume.

Mass Scientists measure mass in grams. You might use a beam balance similar to the one shown in **Figure 13.** The balance has a pan on one side and a set of beams on the other side. Each beam has a rider that slides on the beam.

Before you find the mass of an object, slide all the riders back to the zero point. Check the pointer on the right to make sure it swings an equal distance above and below the zero point. If the swing is unequal, find and turn the adjusting screw until you have an equal swing.

Place an object on the pan. Slide the largest rider along its beam until the pointer drops below zero. Then move it back one notch. Repeat the process on each beam until the pointer swings an equal distance above and below the zero point. Sum the masses on each beam to find the mass of the object. Move all riders back to zero when finished.

Figure 13
A triple beam balance is used to determine the mass of an object.

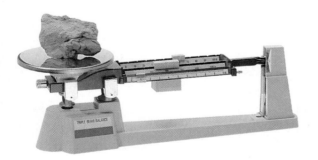

You should never place a hot object on the pan or pour chemicals directly onto the pan. Instead, find the mass of a clean container. Remove the container from the pan, then place the chemicals in the container. Find the mass of the container with the chemicals in it. To find the mass of the chemicals, subtract the mass of the empty container from the mass of the filled container.

Making and Using Tables

Browse through your textbook and you will see tables in the text and in the activities. In a table, data, or information, are arranged so that they are easier to understand. Activity tables help organize the data you collect during an activity so results can be interpreted.

Making Tables To make a table, list the items to be compared in the first column and the characteristics to be compared in the first row. The title should clearly indicate the content of the table, and the column or row heads should tell the reader what information is found in there. The table below lists materials collected for recycling on three weekly pick-up days. The inclusion of kilograms in parentheses also identifies for the reader that the figures are mass units.

Recyclable Materials Collected During Week			
Day of Week	Paper (kg)	Aluminum (kg)	Glass (kg)
Monday	5.0	4.0	12.0
Wednesday	4.0	1.0	10.0
Friday	2.5	2.0	10.0

Using Tables How much paper, in kilograms, is being recycled on Wednesday? Locate the column labeled "Paper (kg)" and the row "Wednesday." The information in the box where the column and row intersect is the answer. Did you answer "4.0"? How much aluminum, in kilograms, is being recycled on Friday? If you answered "2.0," you understand how to read the table. How much glass is collected for recycling each week? Locate the column labeled "Glass (kg)" and add the figures for all three rows. If you answered "32.0," then you know how to locate and use the data provided in the table.

Recording Data

To be useful, the data you collect must be recorded carefully. Accuracy is key. A well-thought-out experiment includes a way to record procedures, observations, and results accurately. Data tables are one way to organize and record results. Set up the tables you will need ahead of time so you can record the data right away.

Record information properly and neatly. Never put unidentified data on scraps of paper. Instead, data should be written in a notebook like the one in **Figure 14.** Write in pencil so information isn't lost if your data gets wet. At each point in the experiment, record your data and label it. That way, your information will be accurate and you will not have to determine what the figures mean when you look at your notes later.

Figure 14
Record data neatly and clearly so it is easy to understand.

Recording Observations

It is important to record observations accurately and completely. That is why you always should record observations in your notes immediately as you make them. It is easy to miss details or make mistakes when recording results from memory. Do not include your personal thoughts when you record your data. Record only what you observe to eliminate bias. For example, when you record that a plant grew 12 cm in one day, you would note that this was the largest daily growth for the week. However, you would not refer to the data as "the best growth spurt of the week."

Making Models

You can organize the observations and other data you collect and record in many ways. Making models is one way to help you better understand the parts of a structure you have been observing or the way a process for which you have been taking various measurements works.

Models often show things that are very large or small or otherwise would be difficult to see and understand. You can study blood vessels and know that they are hollow tubes. The size and proportional differences among arteries, veins, and capillaries can be explained in words. However, you can better visualize the relative sizes and proportions of blood vessels by making models of them. Gluing different kinds of pasta to thick paper so the openings can be seen can help you see how the differences in size, wall thickness, and shape among types of blood vessels affect their functions.

Other models can be devised on a computer. Some models, such as disease control models used by doctors to predict the spread of the flu, are mathematical and are represented by equations.

Making and Using Graphs

After scientists organize data in tables, they might display the data in a graph that shows the relationship of one variable to another. A graph makes interpretation and analysis of data easier. Three types of graphs are the line graph, the bar graph, and the circle graph.

Line Graphs A line graph like in **Figure 15** is used to show the relationship between two variables. The variables being compared go on two axes of the graph. For data from an experiment, the independent variable always goes on the horizontal axis, called the *x*-axis. The dependent variable always goes on the vertical axis, called the *y*-axis. After drawing your axes, label each with a scale. Next, plot the data points.

A data point is the intersection of the recorded value of the dependent variable for each tested value of the independent variable. After all the points are plotted, connect them.

Bar Graphs Bar graphs compare data that do not change continuously. Vertical bars show the relationships among data.

To make a bar graph, set up the *y*-axis as you did for the line graph. Draw vertical bars of equal size from the *x*-axis up to the point on the *y*-axis that represents the value of *x*.

Figure 16
The number of wing vibrations per second for different insects can be shown as a bar graph or circle graph.

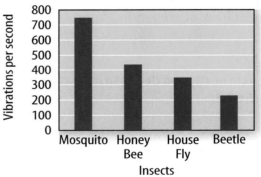

Circle Graphs A circle graph uses a circle divided into sections to display data as parts (fractions or percentages) of a whole. The size of each section corresponds to the fraction or percentage of the data that the section represents. So, the entire circle represents 100 percent, one-half represents 50 percent, one-fifth represents 20 percent, and so on.

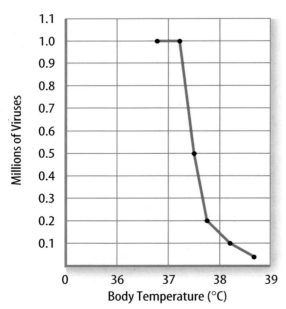

Figure 15
This line graph shows the relationship between body temperature and the millions of infecting viruses present in a human body.

Analyzing and Applying Results

Analyzing Results

To determine the meaning of your observations and investigation results, you will need to look for patterns in the data. You can organize your information in several of the ways that are discussed in this handbook. Then you must think critically to determine what the data mean. Scientists use several approaches when they analyze the data they have collected and recorded. Each approach is useful for identifying specific patterns in the data.

Forming Operational Definitions

An operational definition defines an object by showing how it functions, works, or behaves. Such definitions are written in terms of how an object works or how it can be used; that is, they describe its job or purpose.

For example, a ruler can be defined as a tool that measures the length of an object (how it can be used). A ruler also can be defined as something that contains a series of marks that can be used as a standard when measuring (how it works).

Classifying

Classifying is the process of sorting objects or events into groups based on common features. When classifying, first observe the objects or events to be classified. Then select one feature that is shared by some members in the group but not by all. Place those members that share that feature into a subgroup. You can classify members into smaller and smaller subgroups based on characteristics.

How might you classify a group of animals? You might first classify them by putting all of the dogs, cats, lizards, snakes, and birds into separate groups. Within each group,

you could then look for another common feature by which to further classify members of the group, such as size or color.

Remember that when you classify, you are grouping objects or events for a purpose. For example, classifying animals can be the first step in identifying them. You might know that a cardinal is a red bird. To find it in a large group of animals, you might start with the classification scheme mentioned here. You'll locate a cardinal within the red grouping of the birds that you separate from the rest of the animals. A male ruby-throated hummingbird could be located within the birds by its tiny size and the bright red color of its throat. Keep your purpose in mind as you select the features to form groups and subgroups.

Figure 17
Color is one of many characteristics that are used to classify animals.

Comparing and Contrasting

Observations can be analyzed by noting the similarities and differences between two or more objects or events that you observe. When you look at objects or events to see how they are similar, you are comparing them. Contrasting is looking for differences in objects or events. The table below compares and contrasts the nutritional value of two cereals.

Nutritional Values		
	Cereal A	**Cereal B**
Calories	220	160
Fat	10 g	10 g
Protein	2.5 g	2.6 g
Carbohydrate	30 g	15 g

Recognizing Cause and Effect

Have you ever gotten a cold and then suggested that you probably caught it from a classmate who had one recently? If so, you have observed an effect and inferred a cause. The event is the effect, and the reason for the event is the cause.

When scientists are unsure of the cause of a certain event, they design controlled experiments to determine what caused it.

Interpreting Data

The word *interpret* means "to explain the meaning of something." Look at the problem originally being explored in an experiment and figure out what the data show. Identify the control group and the test group so you can see whether or not changes in the independent variable have had an effect. Look for differences in the dependent variable between the control and test groups.

These differences you observe can be qualitative or quantitative. You would be able to describe a qualitative difference using only words, whereas you would measure a quantitative difference and describe it using numbers. If there are qualitative or quantitative differences, the independent variable that is being tested could have had an effect. If no qualitative or quantitative differences are found between the control and test groups, the variable that is being tested apparently had no effect.

For example, suppose that three pepper plants are placed in a garden and two of the plants are fertilized, but the third is left to grow without fertilizer. Suppose you are then asked to describe any differences in the plants after two weeks. A qualitative difference might be the appearance of brighter green leaves on fertilized plants but not on the unfertilized plant. A quantitative difference might be a difference in the height of the plants or the number of flowers on them.

Inferring Scientists often make inferences based on their observations. An inference is an attempt to explain, or interpret, observations or to indicate what caused what you observed. An inference is a type of conclusion.

When making an inference, be certain to use accurate data and accurately described observations. Analyze all of the data that you've collected. Then, based on everything you know, explain or interpret what you've observed.

Drawing Conclusions

When scientists have analyzed the data they collected, they proceed to draw conclusions about what the data mean. These conclusions are sometimes stated using words similar to those found in the hypothesis formed earlier in the process.

Conclusions To analyze your data, you must review all of the observations and measurements that you made and recorded. Recheck all data for accuracy. After your data are rechecked and organized, you are almost ready to draw a conclusion such as "Plants need sunlight in order to grow."

Before you can draw a conclusion, however, you must determine whether the data allow you to come to a conclusion that supports a hypothesis. Sometimes that will be the case; other times it will not.

If your data do not support a hypothesis, it does not mean that the hypothesis is wrong. It means only that the results of the investigation did not support the hypothesis. Maybe the experiment needs to be redesigned, but very likely, some of the initial observations on which the hypothesis was based were incomplete or biased. Perhaps more observation or research is needed to refine the hypothesis.

Avoiding Bias Sometimes drawing a conclusion involves making judgments. When you make a judgment, you form an opinion about what your data mean. It is important to be honest and to avoid reaching a conclusion if no supporting evidence for it exists or if it is based on a small sample. It also is important not to allow any expectations of results to bias your judgments. If possible, it is a good idea to collect additional data. Scientists do this all the time.

For example, animal behaviorist Katharine Payne made an important observation about elephant communication. While visiting a zoo, Payne felt the air vibrating around her. At the same time, she also noticed that the skin on an elephant's forehead was fluttering. She suspected that the elephants were generating the vibrations and that they might be using the low-frequency sounds to communicate.

Payne conducted an experiment to record these sounds and simultaneously observe the behavior of the elephants in the zoo. She later conducted a similar experiment in Namibia in southwest Africa, where elephant herds roam. The additional data she collected supported the judgment Payne had made, which was that these low-frequency sounds were a form of communication between elephants.

Evaluating Others' Data and Conclusions

Sometimes scientists have to use data that they did not collect themselves, or they have to rely on observations and conclusions drawn by other researchers. In cases such as these, the data must be evaluated carefully.

How were the data obtained? How was the investigation done? Has it been duplicated by other researchers? Did they come up with the same results? Look at the conclusion, as well. Would you reach the same conclusion from these results? Only when you have confidence in the data of others can you believe it is true and feel comfortable using it.

Communicating

The communication of ideas is an important part of the work of scientists. A discovery that is not reported will not advance the scientific community's understanding or knowledge. Communication among scientists also is important as a way of improving their investigations.

Scientists communicate in many ways, from writing articles in journals and magazines that explain their investigations and experiments, to announcing important discoveries on television and radio, to sharing ideas with colleagues on the Internet or presenting them as lectures.

People who study science rely on computers to record and store data and to analyze results from investigations. Whether you work in a laboratory or just need to write a lab report with tables, good computer skills are a necessity.

Using a Word Processor

Suppose your teacher has assigned a written report. After you've completed your research and decided how you want to write the information, you need to put all that information on paper. The easiest way to do this is with a word processing application on a computer.

A computer application that allows you to type your information, change it as many times as you need to, and then print it out so that it looks neat and clean is called a word processing application. You also can use this type of application to create tables and columns, add bullets or cartoon art to your page, include page numbers, and even check your spelling.

Helpful Hints

- If you aren't sure how to do something using your word processing program, look in the help menu. You will find a list of topics there to click on for help. After you locate the help topic you need, just follow the step-by-step instructions you see on your screen.
- Just because you've spell checked your report doesn't mean that the spelling is perfect. The spell check feature can't catch misspelled words that look like other words. If you've accidentally typed *wind* instead of *wing*, the spell checker won't know the difference. Always reread your report to make sure you didn't miss any mistakes.

Figure 18
You can use computer programs to make graphs and tables.

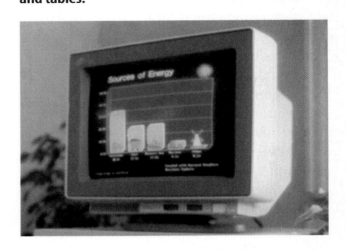

Using a Database

Imagine you're in the middle of a research project busily gathering facts and information. You soon realize that it's becoming more difficult to organize and keep track of all the information. The tool to use to solve information overload is a database. Just as a file cabinet organizes paper records, a database organizes computer records. However, a database is more powerful than a simple file cabinet because at the click of a mouse, the contents can be reshuffled and reorganized. At computer-quick speeds, databases can sort information by any characteristics and filter data into multiple categories.

Helpful Hints

- Before setting up a database, take some time to learn the features of your database software by practicing with established database software.
- Periodically save your database as you enter data. That way, if something happens such as your computer malfunctions or the power goes off, you won't lose all of your work.

Doing a Database Search

When searching for information in a database, use the following search strategies to get the best results. These are the same search methods used for searching Internet databases.

- Place the word *and* between two words in your search if you want the database to look for any entries that have both words. For example, "fox *and* mink" would give you information that mentions both fox and mink.
- Place the word *or* between two words if you want the database to show entries that have at least one of the words. For example "fox *or* mink" would show you information that mentions either fox or mink.
- Place the word *not* between two words if you want the database to look for entries that have the first word but do not have the second word. For example, "canine *not* fox" would show you information that mentions the term *canine* but does not mention the fox.

In summary, databases can be used to store large amounts of information about a particular subject. Databases allow biologists, Earth scientists, and physical scientists to search for information quickly and accurately.

Using an Electronic Spreadsheet

Your science fair experiment has produced lots of numbers. How do you keep track of all the data, and how can you easily work out all the calculations needed? You can use a computer program called a spreadsheet to record data that involve numbers. A spreadsheet is an electronic mathematical worksheet.

Type in your data in rows and columns, just as in a data table on a sheet of paper. A spreadsheet uses simple math to do data calculations. For example, you could add, subtract, divide, or multiply any of the values in the spreadsheet by another number. You also could set up a series of math steps you want to apply to the data. If you want to add 12 to all the numbers and then multiply all the numbers by 10, the computer does all the calculations for you in the spreadsheet. Below is an example of a spreadsheet that records data from an experiment with mice in a maze.

Helpful Hints

- Before you set up the spreadsheet, identify how you want to organize the data. Include any formulas you will need to use.
- Make sure you have entered the correct data into the correct rows and columns.
- You also can display your results in a graph. Pick the style of graph that best represents the data with which you are working.

Figure 19
A spreadsheet allows you to display large amounts of data and do calculations automatically.

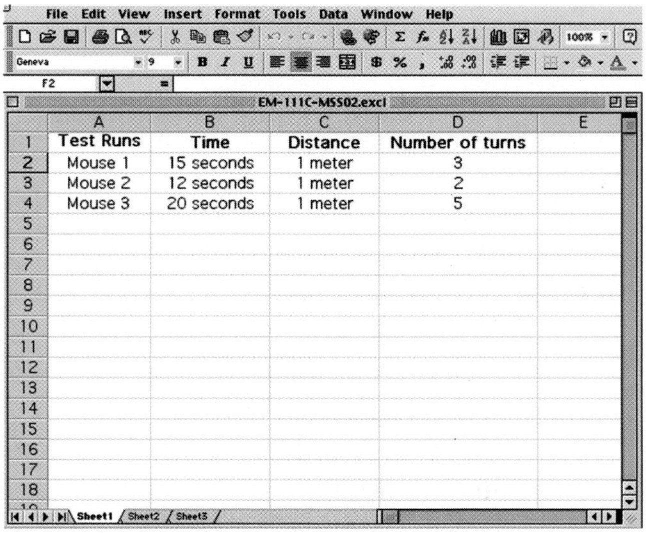

Test Runs	Time	Distance	Number of turns
Mouse 1	15 seconds	1 meter	3
Mouse 2	12 seconds	1 meter	2
Mouse 3	20 seconds	1 meter	5

Using a Computerized Card Catalog

When you have a report or paper to research, you probably go to the library. To find the information you need in the library, you might have to use a computerized card catalog. This type of card catalog allows you to search for information by subject, by title, or by author. The computer then will display all the holdings the library has on the subject, title, or author requested.

A library's holdings can include books, magazines, databases, videos, and audio materials. When you have chosen something from this list, the computer will show whether an item is available and where in the library to find it.

Helpful Hints

- Remember that you can use the computer to search by subject, author, or title. If you know a book's author but not the title, you can search for all the books the library has by that author.
- When searching by subject, it's often most helpful to narrow your search by using specific search terms, such as *and*, *or*, and *not*. If you don't find enough sources, you can broaden your search.
- Pay attention to the type of materials found in your search. If you need a book, you can eliminate any videos or other resources that come up in your search.
- Knowing how your library is arranged can save you a lot of time. The librarian will show you where certain types of materials are kept and how to find specific holdings.

Using Graphics Software

Are you having trouble finding that exact piece of art you're looking for? Do you have a picture in your mind of what you want but can't seem to find the right graphic to represent your ideas? To solve these problems, you can use graphics software. Graphics software allows you to create and change images and diagrams in almost unlimited ways. Typical uses for graphics software include arranging clip art, changing scanned images, and constructing pictures from scratch. Most graphics software applications work in similar ways. They use the same basic tools and functions. Once you master one graphics application, you can use any other graphics application relatively easily.

Figure 20
Graphics software can use your data to draw bar graphs.

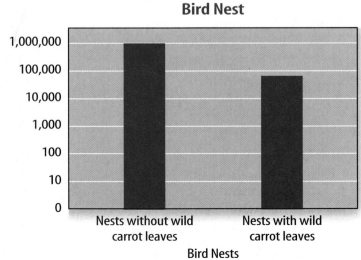

Number of Mites per Bird Nest

Figure 21
Graphics software can use your data to draw circle graphs.

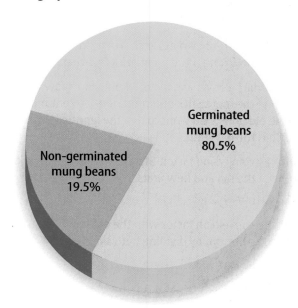

Germinated mung beans 80.5%

Non-germinated mung beans 19.5%

Helpful Hints

- As with any method of drawing, the more you practice using the graphics software, the better your results will be.
- Start by using the software to manipulate existing drawings. Once you master this, making your own illustrations will be easier.
- Clip art is available on CD-ROMs and the Internet. With these resources, finding a piece of clip art to suit your purposes is simple.
- As you work on a drawing, save it often.

Developing Multimedia Presentations

It's your turn—you have to present your science report to the entire class. How do you do it? You can use many different sources of information to get the class excited about your presentation. Posters, videos, photographs, sound, computers, and the Internet can help show your ideas.

First, determine what important points you want to make in your presentation. Then, write an outline of what materials and types of media would best illustrate those points. Maybe you could start with an outline on an overhead projector, then show a video, followed by something from the Internet or a slide show accompanied by music or recorded voices. You might choose to use a presentation builder computer application that can combine all these elements into one presentation. Make sure the presentation is well constructed to make the most impact on the audience.

Figure 22
Multimedia presentations use many types of print and electronic materials.

Helpful Hints

- Carefully consider what media will best communicate the point you are trying to make.
- Make sure you know how to use any equipment you will be using in your presentation.
- Practice the presentation several times.
- If possible, set up all of the equipment ahead of time. Make sure everything is working correctly.

Math Skill Handbook

Use this Math Skill Handbook to help solve problems you are given in this text. You might find it useful to review topics in this Math Skill Handbook first.

Converting Units

In science, quantities such as length, mass, and time sometimes are measured using different units. Suppose you want to know how many miles are in 12.7 km.

Conversion factors are used to change from one unit of measure to another. A conversion factor is a ratio that is equal to one. For example, there are 1,000 mL in 1 L, so 1,000 mL equals 1 L, or:

$$1,000 \text{ mL} = 1 \text{ L}$$

If both sides are divided by 1 L, this equation becomes:

$$\frac{1,000 \text{ mL}}{1 \text{ L}} = 1$$

The **ratio** on the left side of this equation is equal to 1 and is a conversion factor. You can make another conversion factor by dividing both sides of the top equation by 1,000 mL:

$$1 = \frac{1 \text{ L}}{1,000 \text{ mL}}$$

To **convert units,** you multiply by the appropriate conversion factor. For example, how many milliliters are in 1.255 L? To convert 1.255 L to milliliters, multiply 1.255 L by a conversion factor.

Use the **conversion factor** with new units (mL) in the numerator and the old units (L) in the denominator.

$$1.255 \text{ L} \times \frac{1,000 \text{ mL}}{1 \text{ L}} = 1,255 \text{ mL}$$

The unit L divides in this equation, just as if it were a number.

Example 1 There are 2.54 cm in 1 inch. If a meterstick has a length of 100 cm, how long is the meterstick in inches?

Step 1 Decide which conversion factor to use. You know the length of the meterstick in centimeters, so centimeters are the old units. You want to find the length in inches, so inch is the new unit.

Step 2 Form the conversion factor. Start with the relationship between the old and new units.

$$2.54 \text{ cm} = 1 \text{ inch}$$

Step 3 Form the conversion factor with the old unit (centimeter) on the bottom by dividing both sides by 2.54 cm.

$$1 = \frac{2.54 \text{ cm}}{2.54 \text{ cm}} = \frac{1 \text{ inch}}{2.54 \text{ cm}}$$

Step 4 Multiply the old measurement by the conversion factor.

$$100 \text{ cm} \times \frac{1 \text{ inch}}{2.54 \text{ cm}} = 39.37 \text{ inches}$$

The meterstick is 39.37 inches long.

Example 2 There are 365 days in one year. If a person is 14 years old, what is his or her age in days? (Ignore leap years)

Step 1 Decide which conversion factor to use. You want to convert years to days.

Step 2 Form the conversion factor. Start with the relation between the old and new units.

$$1 \text{ year} = 365 \text{ days}$$

Step 3 Form the conversion factor with the old unit (year) on the bottom by dividing both sides by 1 year.

$$1 = \frac{1 \text{ year}}{1 \text{ year}} = \frac{365 \text{ days}}{1 \text{ year}}$$

Step 4 Multiply the old measurement by the conversion factor:

$$14 \text{ years} \times \frac{365 \text{ days}}{1 \text{ year}} = 5,110 \text{ days}$$

The person's age is 5,110 days.

Practice Problem A cat has a mass of 2.31 kg. If there are 1,000 g in 1 kg, what is the mass of the cat in grams?

Using Fractions

A **fraction** is a number that compares a part to the whole. For example, in the fraction $\frac{2}{3}$, the 2 represents the part and the 3 represents the whole. In the fraction $\frac{2}{3}$, the top number, 2, is called the numerator. The bottom number, 3, is called the denominator.

Sometimes fractions are not written in their simplest form. To determine a fraction's **simplest form,** you must find the greatest common factor (GCF) of the numerator and denominator. The greatest common factor is the largest common factor of all the factors the two numbers have in common.

For example, because the number 3 divides into 12 and 30 evenly, it is a common factor of 12 and 30. However, because the number 6 is the largest number that evenly divides into 12 and 30, it is the **greatest common factor.**

After you find the greatest common factor, you can write a fraction in its simplest form. Divide both the numerator and the denominator by the greatest common factor. The number that results is the fraction in its **simplest form.**

Example Twelve of the 20 corn plants in a field are more than 1.5 m tall. What fraction of the corn plants in the field is 1.5 m tall?

Step 1 Write the fraction.

$$\frac{part}{whole} = \frac{12}{20}$$

Step 2 To find the GCF of the numerator and denominator, list all of the factors of each number.

Factors of 12: 1, 2, 3, 4, 6, 12 (the numbers that divide evenly into 12)

Factors of 20: 1, 2, 4, 5, 10, 20 (the numbers that divide evenly into 20)

Step 3 List the common factors.

1, 2, 4.

Step 4 Choose the greatest factor in the list of common factors.

The GCF of 12 and 20 is 4.

Step 5 Divide the numerator and denominator by the GCF.

$$\frac{12 \div 4}{20 \div 4} = \frac{3}{5}$$

In the field, $\frac{3}{5}$ of the corn plants are more than 1.5 m tall.

Practice Problem There are 90 duck eggs in a population. Of those eggs, 66 hatch over a one-week period. What fraction of the eggs hatch over a one-week period? Write the fraction in simplest form.

Math Skill Handbook

Calculating Ratios

A **ratio** is a comparison of two numbers by division.

Ratios can be written 3 to 5 or 3:5. Ratios also can be written as fractions, such as $\frac{3}{5}$. Ratios, like fractions, can be written in simplest form. Recall that a fraction is in **simplest form** when the greatest common factor (GCF) of the numerator and denominator is 1.

Example From a package of sunflower seeds, 40 seeds germinated and 64 did not. What is the ratio of germinated to not germinated seeds as a fraction in simplest form?

Step 1 Write the ratio as a fraction.

$$\frac{\text{germinated}}{\text{not germinated}} = \frac{40}{64}$$

Step 2 Express the fraction in simplest form. The GCF of 40 and 64 is 8.

$$\frac{40}{64} = \frac{40 \div 8}{64 \div 8} = \frac{5}{8}$$

The ratio of germinated to not germinated seeds is $\frac{5}{8}$.

Practice Problem Two children measure 100 cm and 144 cm in height. What is the ratio of their heights in simplest fraction form?

Using Decimals

A **decimal** is a fraction with a denominator of 10, 100, 1,000, or another power of 10. For example, 0.854 is the same as the fraction $\frac{854}{1,000}$.

In a decimal, the decimal point separates the ones place and the tenths place. For example, 0.27 means twenty-seven hundredths, or $\frac{27}{100}$, where 27 is the **number of units** out of 100 units. Any fraction can be written as a decimal using division.

Example Write $\frac{5}{8}$ as a decimal.

Step 1 Write a division problem with the numerator, 5, as the dividend and the denominator, 8, as the divisor. Write 5 as 5.000.

Step 2 Solve the problem.

$$\begin{array}{r} 0.625 \\ 8\overline{)5.000} \\ \underline{4\,8} \\ 20 \\ \underline{16} \\ 40 \\ \underline{40} \\ 0 \end{array}$$

Therefore, $\frac{5}{8} = 0.625$.

Practice Problem Write $\frac{19}{25}$ as a decimal.

Using Percentages

The word *percent* means "out of one hundred." A **percent** is a ratio that compares a number to 100. Suppose you read that 77 percent of all fish on Earth live in the Pacific Ocean. That is the same as reading that the ratio of Earth's fish that live in the Pacific Ocean is $\frac{77}{100}$. To express a fraction as a percent, first find an equivalent decimal for the fraction. Then, multiply the decimal by 100 and add the percent symbol. For example, $\frac{1}{2} = 1 \div 2 = 0.5$. Then $0.5 \cdot 100 = 50 = 50\%$.

Example Express $\frac{13}{20}$ as a percent.

Step 1 Find the equivalent decimal for the fraction.

$$
\begin{array}{r}
0.65 \\
20)\overline{13.00} \\
\underline{120} \\
100 \\
\underline{100} \\
0
\end{array}
$$

Step 2 Rewrite the fraction $\frac{13}{20}$ as 0.65.

Step 3 Multiply 0.65 by 100 and add the % sign.

$0.65 \cdot 100 = 65 = 65\%$

So, $\frac{13}{20} = 65\%$.

Practice Problem In an experimental population of 365 sheep, 73 were brown. What percent of the sheep were brown?

Using Precision and Significant Digits

When you make a **measurement,** the value you record depends on the precision of the measuring instrument. When adding or subtracting numbers with different precision, the answer is rounded to the smallest number of decimal places of any number in the sum or difference. When multiplying or dividing, the answer is rounded to the smallest number of significant figures of any number being multiplied or divided. When counting the number of **significant figures,** all digits are counted except zeros at the end of a number with no decimal such as 2,500, and zeros at the beginning of a decimal such as 0.03020.

Example The lengths 5.28 and 5.2 are measured in meters. Find the sum of these lengths and report the sum using the least precise measurement.

Step 1 Find the sum.

5.28 m	2 digits after the decimal
+ 5.2 m	1 digit after the decimal
10.48 m	

Step 2 Round to one digit after the decimal because the least number of digits after the decimal of the numbers being added is 1.

The sum is 10.5 m.

Practice Problem Multiply the numbers in the example using the rule for multiplying and dividing. Report the answer with the correct number of significant figures.

Solving One-Step Equations

An **equation** is a statement that two things are equal. For example, $A = B$ is an equation that states that A is equal to B.

Sometimes one side of the equation will contain a **variable** whose value is not known. In the equation $3x = 12$, the variable is x.

The equation is solved when the variable is replaced with a value that makes both sides of the equation equal to each other. For example, the solution of the equation $3x = 12$ is $x = 4$. If the x is replaced with 4, then the equation becomes $3 \cdot 4 = 12$, or $12 = 12$.

To solve an equation such as $8x = 40$, divide both sides of the equation by the number that multiplies the variable.

$$8x = 40$$
$$\frac{8x}{8} = \frac{40}{8}$$
$$x = 5$$

You can check your answer by replacing the variable with your solution and seeing if both sides of the equation are the same.

$$8x = 8 \cdot 5 = 40$$

The left and right sides of the equation are the same, so $x = 5$ is the solution.

Sometimes an equation is written in this way: $a = bc$. This also is called a **formula.** The letters can be replaced by numbers, but the numbers must still make both sides of the equation the same.

Example 1 Solve the equation $10x = 35$.

Step 1 Find the solution by dividing each side of the equation by 10.

$$10x = 35 \qquad \frac{10x}{10} = \frac{35}{10} \qquad x = 3.5$$

Step 2 Check the solution.

$$10x = 35 \qquad 10 \times 3.5 = 35 \qquad 35 = 35$$

Both sides of the equation are equal, so $x = 3.5$ is the solution to the equation.

Example 2 In the formula $a = bc$, find the value of c if $a = 20$ and $b = 2$.

Step 1 Rearrange the formula so the unknown value is by itself on one side of the equation by dividing both sides by b.

$$a = bc$$
$$\frac{a}{b} = \frac{bc}{b}$$
$$\frac{a}{b} = c$$

Step 2 Replace the variables a and b with the values that are given.

$$\frac{a}{b} = c$$
$$\frac{20}{2} = c$$
$$10 = c$$

Step 3 Check the solution.

$$a = bc$$
$$20 = 2 \times 10$$
$$20 = 20$$

Both sides of the equation are equal, so $c = 10$ is the solution when $a = 20$ and $b = 2$.

Practice Problem In the formula $h = gd$, find the value of d if $g = 12.3$ and $h = 17.4$.

A **proportion** is an equation that shows that two ratios are equivalent. The ratios $\frac{2}{4}$ and $\frac{5}{10}$ are equivalent, so they can be written as $\frac{2}{4} = \frac{5}{10}$. This equation is an example of a proportion.

When two ratios form a proportion, the **cross products** are equal. To find the cross products in the proportion $\frac{2}{4} = \frac{5}{10}$, multiply the 2 and the 10, and the 4 and the 5. Therefore $2 \cdot 10 = 4 \cdot 5$, or $20 = 20$.

Because you know that both proportions are equal, you can use cross products to find a missing term in a proportion. This is known as **solving the proportion.** Solving a proportion is similar to solving an equation.

Example The heights of a tree and a pole are proportional to the lengths of their shadows. The tree casts a shadow of 24 m at the same time that a 6-m pole casts a shadow of 4 m. What is the height of the tree?

Step 1 Write a proportion.

$$\frac{\text{height of tree}}{\text{height of pole}} = \frac{\text{length of tree's shadow}}{\text{length of pole's shadow}}$$

Step 2 Substitute the known values into the proportion. Let h represent the unknown value, the height of the tree.

$$\frac{h}{6} = \frac{24}{4}$$

Step 3 Find the cross products.

$$h \cdot 4 = 6 \cdot 24$$

Step 4 Simplify the equation.

$$4h = 144$$

Step 5 Divide each side by 4.

$$\frac{4h}{4} = \frac{144}{4}$$

$$h = 36$$

The height of the tree is 36 m.

Practice Problem The proportions of bluefish are stable by the time they reach a length of 30 cm. The distance from the tip of the mouth to the back edge of the gill cover in a 35-cm bluefish is 15 cm. What is the distance from the tip of the mouth to the back edge of the gill cover in a 59-cm bluefish?

Math Skill Handbook

Statistics is the branch of mathematics that deals with collecting, analyzing, and presenting data. In statistics, there are three common ways to summarize the data with a single number—the mean, the median, and the mode.

The **mean** of a set of data is the arithmetic average. It is found by adding the numbers in the data set and dividing by the number of items in the set.

The **median** is the middle number in a set of data when the data are arranged in numerical order. If there were an even number of data points, the median would be the mean of the two middle numbers.

The **mode** of a set of data is the number or item that appears most often.

Another number that often is used to describe a set of data is the range. The **range** is the difference between the largest number and the smallest number in a set of data.

A **frequency table** shows how many times each piece of data occurs, usually in a survey. The frequency table below shows the results of a student survey on favorite color.

Color	Tally	Frequency
red	||||	4
blue	⌗	5
black	||	2
green	|||	3
purple	⌗ ||	7
yellow	⌗ |	6

Based on the frequency table data, which color is the favorite?

Example The high temperatures (in °C) on five consecutive days in a desert habitat under study are 39°, 37°, 44°, 36°, and 44°. Find the mean, median, mode, and range of this set.

To find the mean:
Step 1 Find the sum of the numbers.

$$39 + 37 + 44 + 36 + 44 = 200$$

Step 2 Divide the sum by the number of items, which is 5.

$$200 \div 5 = 40$$

The mean high temperature is 40°C.

To find the median:
Step 1 Arrange the temperatures from least to greatest.

$$36, \ 37, \ \underline{39}, \ 44, \ 44$$

Step 2 Determine the middle temperature.

The median high temperature is 39°C.

To find the mode:
Step 1 Group the numbers that are the same together.

$$44, 44, 36, 37, 39$$

Step 2 Determine the number that occurs most in the set.

$$\underline{44, 44}, 36, 37, 39$$

The mode measure is 44°C.

To find the range:
Step 1 Arrange the temperatures from largest to smallest.

$$44, 44, 39, 37, 36$$

Step 2 Determine the largest and smallest temperature in the set.

$$\underline{44}, 44, 39, 37, \underline{36}$$

Step 3 Find the difference between the largest and smallest temperatures.

$$44 - 36 = 8$$

The range is 8°C.

Practice Problem Find the mean, median, mode, and range for the data set 8, 4, 12, 8, 11, 14, 16.

Care and Use of a Microscope

Eyepiece Contains magnifying lenses you look through.

Arm Supports the body tube.

Low-power objective Contains the lens with the lowest power magnification.

Stage clips Hold the microscope slide in place.

Fine adjustment Sharpens the image under high magnification.

Coarse adjustment Focuses the image under low power.

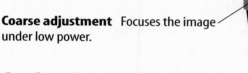

Body tube Connects the eyepiece to the revolving nosepiece.

Revolving nosepiece Holds and turns the objectives into viewing position.

High-power objective Contains the lens with the highest magnification.

Stage Supports the microscope slide.

Light source Provides light that passes upward through the diaphragm, the specimen, and the lenses.

Base Provides support for the microscope.

Caring for a Microscope

1. Always carry the microscope holding the arm with one hand and supporting the base with the other hand.

2. Don't touch the lenses with your fingers.

3. The coarse adjustment knob is used only when looking through the lowest-power objective lens. The fine adjustment knob is used when the high-power objective is in place.

4. Cover the microscope when you store it.

Using a Microscope

1. Place the microscope on a flat surface that is clear of objects. The arm should be toward you.

2. Look through the eyepiece. Adjust the diaphragm so light comes through the opening in the stage.

3. Place a slide on the stage so the specimen is in the field of view. Hold it firmly in place by using the stage clips.

4. Always focus with the coarse adjustment and the low-power objective lens first. After the object is in focus on low power, turn the nosepiece until the high-power objective is in place. Use ONLY the fine adjustment to focus with the high-power objective lens.

Making a Wet-Mount Slide

1. Carefully place the item you want to look at in the center of a clean, glass slide. Make sure the sample is thin enough for light to pass through.

2. Use a dropper to place one or two drops of water on the sample.

3. Hold a clean coverslip by the edges and place it at one edge of the water. Slowly lower the coverslip onto the water until it lies flat.

4. If you have too much water or a lot of air bubbles, touch the edge of a paper towel to the edge of the coverslip to draw off extra water and draw out unwanted air.

Safety in the Science Classroom

1. Always obtain your teacher's permission to begin an investigation.

2. Study the procedure. If you have questions, ask your teacher. Be sure you understand any safety symbols shown on the page.

3. Use the safety equipment provided for you. Goggles and a safety apron should be worn during most investigations.

4. Always slant test tubes away from yourself and others when heating them or adding substances to them.

5. Never eat or drink in the lab, and never use lab glassware as food or drink containers. Never inhale chemicals. Do not taste any substances or draw any material into a tube with your mouth.

6. Report any spill, accident, or injury, no matter how small, immediately to your teacher, then follow his or her instructions.

7. Know the location and proper use of the fire extinguisher, safety shower, fire blanket, first aid kit, and fire alarm.

8. Keep all materials away from open flames. Tie back long hair and tie down loose clothing.

9. If your clothing should catch fire, smother it with the fire blanket, or get under a safety shower. NEVER RUN.

10. If a fire should occur, turn off the gas then leave the room according to established procedures.

Follow these procedures as you clean up your work area

1. Turn off the water and gas. Disconnect electrical devices.

2. Clean all pieces of equipment and return all materials to their proper places.

3. Dispose of chemicals and other materials as directed by your teacher. Place broken glass and solid substances in the proper containers. Make sure never to discard materials in the sink.

4. Clean your work area. Wash your hands thoroughly after working in the laboratory.

First Aid	
Injury	**Safe Response ALWAYS NOTIFY YOUR TEACHER IMMEDIATELY**
Burns	Apply cold water.
Cuts and Bruises	Stop any bleeding by applying direct pressure. Cover cuts with a clean dressing. Apply ice packs or cold compresses to bruises.
Fainting	Leave the person lying down. Loosen any tight clothing and keep crowds away.
Foreign Matter in Eye	Flush with plenty of water. Use eyewash bottle or fountain.
Poisoning	Note the suspected poisoning agent.
Any Spills on Skin	Flush with large amounts of water or use safety shower.

REFERENCE HANDBOOK C

SI—Metric/English, English/Metric Conversions

	When you want to convert:	To:	Multiply by:
Length	inches	centimeters	2.54
	centimeters	inches	0.39
	yards	meters	0.91
	meters	yards	1.09
	miles	kilometers	1.61
	kilometers	miles	0.62
Mass and Weight*	ounces	grams	28.35
	grams	ounces	0.04
	pounds	kilograms	0.45
	kilograms	pounds	2.2
	tons (short)	tonnes (metric tons)	0.91
	tonnes (metric tons)	tons (short)	1.10
	pounds	newtons	4.45
	newtons	pounds	0.22
Volume	cubic inches	cubic centimeters	16.39
	cubic centimeters	cubic inches	0.06
	liters	quarts	1.06
	quarts	liters	0.95
	gallons	liters	3.78
Area	square inches	square centimeters	6.45
	square centimeters	square inches	0.16
	square yards	square meters	0.83
	square meters	square yards	1.19
	square miles	square kilometers	2.59
	square kilometers	square miles	0.39
	hectares	acres	2.47
	acres	hectares	0.40
Temperature	To convert °Celsius to °Fahrenheit		$°C \times 9/5 + 32$
	To convert °Fahrenheit to °Celsius		$5/9 \ (°F - 32)$

*Weight is measured in standard Earth gravity.

Diversity of Life: Classification of Living Organisms

A six-kingdom system of classification of organisms is used today. Two kingdoms—Kingdom Archaebacteria and Kingdom Eubacteria—contain organisms that do not have a nucleus and that lack membrane-bound structures in the cytoplasm of their cells. The members of the other four kingdoms have a cell or cells that contain a nucleus and structures in the cytoplasm, some of which are surrounded by membranes. These kingdoms are Kingdom Protista, Kingdom Fungi, Kingdom Plantae, and Kingdom Animalia.

Kingdom Archaebacteria

one-celled; some absorb food from their surroundings; some are photosynthetic; some are chemosynthetic; many are found in extremely harsh environments including salt ponds, hot springs, swamps, and deep-sea hydrothermal vents

Kingdom Eubacteria

one-celled; most absorb food from their surroundings; some are photosynthetic; some are chemosynthetic; many are parasites; many are round, spiral, or rod-shaped; some form colonies

Kingdom Protista

Phylum Euglenophyta one-celled; photosynthetic or take in food; most have one flagellum; euglenoids

Phylum Bacillariophyta one-celled; photosynthetic; have unique double shells made of silica; diatoms

Phylum Dinoflagellata one-celled; photosynthetic; contain red pigments; have two flagella; dinoflagellates

Phylum Chlorophyta one-celled, many-celled, or colonies; photosynthetic; contain chlorophyll; live on land, in freshwater, or salt water; green algae

Phylum Rhodophyta most are many-celled; photosynthetic; contain red pigments; most live in deep, saltwater environments; red algae

Phylum Phaeophyta most are many-celled; photosynthetic; contain brown pigments; most live in saltwater environments; brown algae

Phylum Rhizopoda one-celled; take in food; are free-living or parasitic; move by means of pseudopods; amoebas

Kingdom Eubacteria
Bacillus anthracis

Phylum Chlorophyta
Desmids

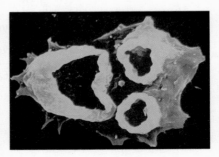

Amoeba

Phylum Zoomastigina one-celled; take in food; free-living or parasitic; have one or more flagella; zoomastigotes

Phylum Ciliophora one-celled; take in food; have large numbers of cilia; ciliates

Phylum Sporozoa one-celled; take in food; have no means of movement; are parasites in animals; sporozoans

Phylum Myxomycota
Slime mold

Phyla Myxomycota and Acrasiomycota one- or many-celled; absorb food; change form during life cycle; cellular and plasmodial slime molds

Phylum Oomycota many-celled; are either parasites or decomposers; live in freshwater or salt water; water molds, rusts and downy mildews

Kingdom Fungi

Phylum Zygomycota many-celled; absorb food; spores are produced in sporangia; zygote fungi; bread mold

Phylum Ascomycota one- and many-celled; absorb food; spores produced in asci; sac fungi; yeast

Phylum Basidiomycota many-celled; absorb food; spores produced in basidia; club fungi; mushrooms

Phylum Deuteromycota members with unknown reproductive structures; imperfect fungi; *Penicillium*

Mycophycota organisms formed by symbiotic relationship between an ascomycote or a basidiomycote and green alga or cyanobacterium; lichens

Phylum Oomycota
Phytophthora infestans

Lichens

Kingdom Plantae

Divisions Bryophyta (mosses), **Anthocerophyta** (hornworts), **Hepatophytal** (liverworts), **Psilophytal** (whisk ferns) many-celled nonvascular plants; reproduce by spores produced in capsules; green; grow in moist, land environments

Division Lycophyta many-celled vascular plants; spores are produced in conelike structures; live on land; are photosynthetic; club mosses

Division Sphenophyta vascular plants; ribbed and jointed stems; scalelike leaves; spores produced in conelike structures; horsetails

Division Pterophyta vascular plants; leaves called fronds; spores produced in clusters of sporangia called sori; live on land or in water; ferns

Division Ginkgophyta deciduous trees; only one living species; have fan-shaped leaves with branching veins and fleshy cones with seeds; ginkgoes

Division Cycadophyta palmlike plants; have large, featherlike leaves; produces seeds in cones; cycads

Division Coniferophyta deciduous or evergreen; trees or shrubs; have needlelike or scalelike leaves; seeds produced in cones; conifers

Division Anthophyta
Tomato plant

Phylum Platyhelminthes
Flatworm

Division Gnetophyta shrubs or woody vines; seeds are produced in cones; division contains only three genera; gnetum

Division Anthophyta dominant group of plants; flowering plants; have fruits with seeds

Kingdom Animalia

Phylum Porifera aquatic organisms that lack true tissues and organs; are asymmetrical and sessile; sponges

Phylum Cnidaria radially symmetrical organisms; have a digestive cavity with one opening; most have tentacles armed with stinging cells; live in aquatic environments singly or in colonies; includes jellyfish, corals, hydra, and sea anemones

Phylum Platyhelminthes bilaterally symmetrical worms; have flattened bodies; digestive system has one opening; parasitic and free-living species; flatworms

Division Bryophyta
Liverwort

Phylum Chordata

Phylum Nematoda round, bilaterally symmetrical body; have digestive system with two openings; free-living forms and parasitic forms; roundworms

Phylum Mollusca soft-bodied animals, many with a hard shell and soft foot or footlike appendage; a mantle covers the soft body; aquatic and terrestrial species; includes clams, snails, squid, and octopuses

Phylum Annelida bilaterally symmetrical worms; have round, segmented bodies; terrestrial and aquatic species; includes earthworms, leeches, and marine polychaetes

Phylum Arthropoda largest animal group; have hard exoskeletons, segmented bodies, and pairs of jointed appendages; land and aquatic species; includes insects, crustaceans, and spiders

Phylum Echinodermata marine organisms; have spiny or leathery skin and a water-vascular system with tube feet; are radially symmetrical; includes sea stars, sand dollars, and sea urchins

Phylum Chordata organisms with internal skeletons and specialized body systems; most have paired appendages; all at some time have a notochord, nerve cord, gill slits, and a postanal tail; include fish, amphibians, reptiles, birds, and mammals

English Glossary

This glossary defines each key term that appears in bold type in the text. It also shows the chapter, section, and page number where you can find the word used.

A

active immunity: long-lasting immunity that results when the body makes its own antibodies in response to a specific antigen. (Chap. 7, Sec. 1, p. 179)

allergen: substance that causes an allergic reaction. (Chap. 7, Sec. 3, p. 191)

allergy: overly strong reaction of the immune system to a foreign substance. (Chap. 7, Sec. 3, p. 190)

alveoli (al VEE uh li): tiny, thin-walled, grapelike clusters at the end of each bronchiole that are surrounded by capillaries, where carbon dioxide and oxygen exchange takes place. (Chap. 4, Sec. 1, p. 95)

amino acid: building block of protein. (Chap. 2, Sec. 1, p. 37)

amniotic (am nee AH tihk) **sac:** thin, liquid-filled, protective membrane that forms around the embryo. (Chap. 6, Sec. 3, p. 159)

antibody: a protein made in response to a specific antigen that can attach to the antigen and cause it to be useless. (Chap. 7, Sec. 1, p. 178)

antigen (AN tih jun): complex molecule that is foreign to your body. (Chap. 7, Sec. 1, p. 178)

artery: blood vessel that carries blood away from the heart and has thick, elastic walls made of connective tissue and smooth muscle tissue. (Chap. 3, Sec. 1, p. 68)

asthma: lung disorder in which the bronchial tubes contract quickly and cause shortness of breath, wheezing, or coughing; may occur as an allergic reaction. (Chap. 4, Sec. 1, p. 100)

atriums (AY tree umz): two upper chambers of the heart that contract at the same time during a heartbeat. (Chap. 3, Sec. 1, p. 65)

axon (AK sahn): neuron structure that carries messages away from the cell body. (Chap. 5, Sec. 1, p. 119)

B

biological vector: disease-carrying organism, such as a rat, mosquito, or fly, that spreads infectious disease. (Chap. 7, Sec. 2, p. 185)

bladder: elastic, muscular organ that holds urine until it leaves the body through the urethra. (Chap. 4, Sec. 2, p. 104)

brain stem: connects the brain to the spinal cord and is made up of the midbrain, the pons, and the medulla. (Chap. 5, Sec. 1, p. 122)

bronchi (BRAHN ki): two short tubes that branch off the lower end of the trachea and carry air into the lungs. (Chap. 4, Sec. 1, p. 95)

C

capillary: microscopic blood vessel that connects arteries and veins; has walls one cell thick, through which nutrients and oxygen diffuse into body cells and waste materials and carbon dioxide diffuse out. (Chap. 3, Sec. 1, p. 69)

carbohydrate (kar boh HI drayt): nutrient that usually is the body's main source of energy. (Chap. 2, Sec. 1, p. 38)

cardiac muscle: striated, involuntary muscle found only in the heart. (Chap. 1, Sec. 2, p. 17)

cartilage: thick, smooth, flexible and slippery tissue layer that covers the ends of bones, makes movement easier by reducing friction, and absorbs shocks. (Chap. 1, Sec. 1, p. 10)

central nervous system: division of the nervous system, made up of the brain and spinal cord. (Chap. 5, Sec. 1, p. 121)

cerebellum (sur uh BEL um): part of the brain that controls voluntary muscle movements, maintains muscle tone, and helps maintain balance. (Chap. 5, Sec. 1, p. 122)

cerebrum (suh REE brum): largest part of the brain, where memory is stored, movements are controlled, and impulses from the senses are interpreted. (Chap. 5, Sec. 1, p. 122)

chemical digestion: occurs when enzymes and other chemicals break down large food molecules into smaller ones. (Chap. 2, Sec. 2, p. 47)

chemotherapy (kee moh THAYR uh pee): use of chemicals to destroy cancer cells. (Chap. 7, Sec. 3, p. 194)

chyme (KIME): liquid product of digestion. (Chap. 2, Sec. 2, p. 51)

cochlea (KOH klee uh): fluid-filled structure in the inner ear in which sound vibrations are converted into nerve impulses that are sent to the brain. (Chap. 5, Sec. 2, p. 132)

coronary (KOR uh ner ee) **circulation:** flow of blood to and from the tissues of the heart. (Chap. 3, Sec. 1, p. 65)

D
---

dendrite: neuron structure that receives messages and sends them to the cell body. (Chap. 5, Sec. 1, p. 119)

dermis: skin layer below the epidermis that contains blood vessels, nerves, oil and sweat glands, and other structures. (Chap. 1, Sec. 3, p. 21)

diaphragm (DI uh fram): muscle beneath the lungs that contracts and relaxes to move gases in and out of the body. (Chap. 4, Sec. 1, p. 96)

digestion: mechanical and chemical breakdown of food into small molecules that cells can absorb and use. (Chap. 2, Sec. 2, p. 47)

E
---

embryo: fertilized egg that has attached to the wall of the uterus. (Chap. 6, Sec. 3, p. 159)

emphysema (em fuh SEE muh): lung disease in which the alveoli enlarge. (Chap. 4, Sec. 1, p. 99)

enzyme: a type of protein that speeds up chemical reactions in the body without being changed or used up itself. (Chap. 2, Sec. 2, p. 48)

epidermis: outer, thinnest skin layer that constantly produces new cells to replace the dead cells rubbed off its surface. (Chap. 1, Sec. 3, p. 20)

F
---

fat: nutrient that stores energy, cushions organs, and helps the body absorb vitamins. (Chap. 2, Sec. 1, p. 39)

fetal stress: can occur during the birth process or after birth as an infant adjusts from a watery, dark, constant-temperature environment to its new environment. (Chap. 6, Sec. 3, p. 162)

fetus: a developing baby after the first two months of pregnancy until birth. (Chap. 6, Sec. 3, p. 160)

food group: group of foods—such as bread, cereal, rice, and pasta—containing the same type of nutrients. (Chap. 2, Sec. 1, p. 44)

English Glossary

H

hemoglobin (HEE muh gloh bun): chemical in red blood cells that carries oxygen from the lungs to body cells and carries some carbon dioxide from body cells back to the lungs. (Chap. 3, Sec. 2, p. 75)

homeostasis: regulation of an organism's internal, life-maintaining conditions despite changes in its environment. (Chap. 5, Sec. 1, p. 119)

hormone (HOR mohn): chemical produced by the endocrine system, released directly into the bloodstream by ductless glands; affects specific target tissues, and can speed up or slow down cellular activities. (Chap. 6, Sec. 1, p. 146)

I

immune system: complex group of defenses that protects the body against pathogens—includes the skin and respiratory, digestive, and circulatory systems. (Chap. 7, Sec. 1, p. 176)

infectious disease: disease caused by a virus, bacterium, fungus, or protist that is spread from an infected organism or the environment to another organism. (Chap. 7, Sec. 2, p. 185)

involuntary muscle: muscle, such as heart muscle, that cannot be consciously controlled. (Chap. 1, Sec. 2, p. 15)

J

joint: any place where two or more bones come together; can be movable or immovable. (Chap. 1, Sec. 1, p. 11)

K

kidney: bean-shaped urinary system organ that is made up of about 1 million nephrons and filters blood, producing urine. (Chap. 4, Sec. 2, p. 102)

L

larynx: airway to which the vocal cords are attached. (Chap. 4, Sec. 1, p. 95)

ligament: tough band of tissue that holds bones together at joints. (Chap. 1, Sec. 1, p. 11)

lymph (LIHMF): tissue fluid that has diffused into the capillaries. (Chap. 3, Sec. 3, p. 80)

lymphatic system: carries lymph though a network of lymph capillaries and vessels and drains it into large veins near the heart; helps fight infections and diseases. (Chap. 3, Sec. 3, p. 80)

lymph node: bean-shaped organ found throughout the body that filters out microorganisms and foreign materials taken up by the lymphocytes. (Chap. 3, Sec. 3, p. 81)

lymphocyte (LIHM fuh site): a type of white blood cell that fights infection. (Chap. 3, Sec. 3, p. 80)

M

mechanical digestion: breakdown of food through chewing, mixing, and churning. (Chap. 2, Sec. 2, p. 47)

melanin: pigment produced by the epidermis that protects skin from sun damage and gives skin and eyes their color. (Chap. 1, Sec. 3, p. 21)

menstrual cycle: hormone-controlled monthly cycle of changes in the female reproductive system that includes the maturation of an egg and preparation of the uterus for possible pregnancy. (Chap. 6, Sec. 2, p. 154)

menstruation (men STRAY shun): monthly flow of blood and tissue cells that occurs when the lining of the uterus breaks down and is shed. (Chap. 6, Sec. 2, p. 154)

mineral: inorganic nutrient that regulates many chemical reactions in the body. (Chap. 2, Sec. 1, p. 42)

muscle: organ that can relax, contract, and provide the force to move bones and body parts. (Chap. 1, Sec. 2, p. 14)

N

nephron (NEF rahn): tiny filtering unit of the kidney. (Chap. 4, Sec. 2, p. 103)

neuron (NOO rahn): basic functioning unit of the nervous system, made up of a cell body, dendrites, and axons. (Chap. 5, Sec. 1, p. 119)

noninfectious disease: disease, such as cancer, diabetes, or asthma, that is not spread from one person to another. (Chap. 7, Sec. 3, p. 190)

nutrients (NEW tree unts): substances in foods—proteins, carbohydrates, fats, vitamins, minerals, and water—that provide energy and materials for cell development, growth, and repair. (Chap. 2, Sec. 1, p. 36)

O

olfactory (ohl FAK tree) **cell:** nasal nerve cell that becomes stimulated by molecules in the air and sends impulses to the brain for interpretation of odors. (Chap. 5, Sec. 2, p. 133)

ovary: female reproductive organ that produces eggs and is located in the lower part of the body. (Chap. 6, Sec. 2, p. 153)

ovulation (ahv yuh LAY shun): monthly process in which an egg is released from an ovary and enters the oviduct, where it can become fertilized by sperm. (Chap. 6, Sec. 2, p. 153)

P

passive immunity: immunity that results when antibodies produced in one animal are introduced into another's body; does not last as long as active immunity. (Chap. 7, Sec. 1, p. 179)

pasteurization (pas chur ruh ZAY shun): process in which a liquid is heated to a temperature that kills most bacteria. (Chap. 7, Sec. 2, p. 181)

periosteum (pur ee AHS tee um): tough, tight-fitting membrane that covers a bone's surface and contains blood vessels that transport nutrients into the bone. (Chap. 1, Sec. 1, p. 9)

peripheral nervous system: division of the nervous system, made up of all the nerves outside the CNS; connects the brain and spinal cord to other body parts. (Chap. 5, Sec. 1, p. 121)

peristalsis (per uh STAHL sus): waves of muscular contractions that move food through the digestive tract. (Chap. 2, Sec. 2, p. 50)

pharynx (FER ingks): tubelike passageway for food, liquid, and air. (Chap. 4, Sec. 1, p. 94)

plasma: liquid part of blood, made mostly of water, in which oxygen, nutrients, and minerals are dissolved. (Chap. 3, Sec. 2, p. 74)

platelet: irregularly shaped cell fragment that helps clot blood and releases chemicals that help form fibrin. (Chap. 3, Sec. 2, p. 75)

pregnancy: period of development—usually about 38 or 39 weeks in humans—from fertilized egg until birth. (Chap. 6, Sec. 3, p. 158)

protein: nutrient made up of amino acids that is used by the body for growth and for replacement and repair of body cells. (Chap. 2, Sec. 1, p. 37)

pulmonary circulation: flow of blood through the heart to the lungs and back to the heart. (Chap. 3, Sec. 1, p. 66)

R

reflex: automatic, involuntary response to a stimulus; controlled by the spinal cord. (Chap. 5, Sec. 1, p. 125)

retina: light-sensitive tissue at the back of the eye; contains rods and cones. (Chap. 5, Sec. 2, p. 129)

S

semen (SEE mun): mixture of sperm and a fluid that helps sperm move and supplies them with an energy source. (Chap. 6, Sec. 2, p. 152)

sexually transmitted disease (STD): infectious disease, such as chlamydia, AIDS, or genital herpes, that is passed from one person to another during sexual contact. (Chap. 7, Sec. 2, p. 186)

skeletal muscle: voluntary, striated muscle that moves bones, works in pairs, and is attached to bones by tendons. (Chap. 1, Sec. 2, p. 17)

skeletal system: all the bones in the body; forms an internal, living framework that provides shape and support, protects internal organs, moves bones, forms blood cells, and stores calcium and phosphorus compounds for later use. (Chap. 1, Sec. 1, p. 8)

smooth muscle: involuntary, nonstriated muscle that controls movement of internal organs. (Chap. 1, Sec. 2, p. 17)

sperm: male reproductive cells produced in the testes. (Chap. 6, Sec. 2, p. 152)

synapse (SIHN aps): small space across which an impulse moves from an axon to the dendrites or cell body of another neuron. (Chap. 5, Sec. 1, p. 121)

systemic circulation: largest part of the circulatory system in which oxygen-rich blood flows to all the organs and body tissues, except the heart and lungs, and oxygen-poor blood is returned to the heart. (Chap. 3, Sec. 1, p. 67)

T

taste bud: major sensory receptor on the tongue; contains taste hairs that send impulses to the brain for interpretation of tastes. (Chap. 5, Sec. 2, p. 134)

tendon: thick band of tissue that attaches bones to muscles. (Chap. 1, Sec. 2, p. 17)

testis: male organ that produces sperm and testosterone. (Chap. 6, Sec. 2, p. 152)

trachea (TRAY kee uh): air-conducting tube that connects the larynx with the bronchi, is lined with mucous membranes and cilia, and contains strong cartilage rings. (Chap. 4, Sec. 1, p. 95)

U

ureter: tube that carries urine from each kidney to the bladder. (Chap. 4, Sec. 2, p. 104)

urethra (yoo REE thruh): tube that carries urine from the bladder to the outside of the body. (Chap. 4, Sec. 2, p. 104)

urinary system: system of excretory organs that rids the blood of wastes, controls blood volume by removing excess water, and balances concentrations of salts and water. (Chap. 4, Sec. 2, p. 101)

urine: wastewater that contains excess water, salts, and other wastes that are not reabsorbed by the body. (Chap. 4, Sec. 2, p. 102)

uterus: hollow, muscular, pear-shaped organ where a fertilized egg develops into a baby. (Chap. 6, Sec. 2, p. 153)

V

vaccination: process of giving a vaccine by mouth or by injection to provide active immunity against a disease. (Chap. 7, Sec. 1, p. 179)

vagina (vuh JI nuh): muscular tube that connects the lower end of the uterus to the outside of the body; the birth canal through which a baby travels when being born. (Chap. 6, Sec. 2, p. 153)

vein: blood vessel that carries blood back to the heart and has one-way valves that keep blood moving toward the heart. (Chap. 3, Sec. 1, p. 68)

ventricles (VEN trih kulz): two lower chambers of the heart that contract at the same time during a heartbeat. (Chap. 3, Sec. 1, p. 65)

villi (VIHL I): fingerlike projections covering the wall of the small intestine that increase the surface area for food absorption. (Chap. 2, Sec. 2, p. 52)

virus: extremely tiny piece of genetic material that infects and multiplies in host cells. (Chap. 7, Sec. 2, p. 182)

vitamin: water-soluble or fat-soluble organic nutrient needed in small quantities for growth, for preventing some diseases, and for regulating body functions. (Chap. 2, Sec. 1, p. 40)

voluntary muscle: muscle, such as a leg or arm muscle, that can be consciously controlled. (Chap. 1, Sec. 2, p. 15)

Spanish Glossary

A

active immunity / inmunidad activa: inmunidad duradera que resulta cuando el cuerpo produce sus propios anticuerpos como una reacción a un antígeno específico. (Cap. 7, Sec. 1, pág. 179)

allergen / alérgeno: sustancia que provoca una reacción alérgica. (Cap. 7, Sec. 3, pág. 191)

allergy / alergia: reacción hipersensible del sistema inmunológico frente a una sustancia extraña. (Cap. 7, Sec. 3, pág. 190)

alveoli / alvéolos: racimos minúsculos de paredes finas que se hallan en el extremo de cada bronquiolo y que están rodeados de capilares, donde se lleva a cabo el intercambio de dióxido de carbono y oxígeno. (Cap. 4, Sec. 1, pág. 95)

amino acids / aminoácidos: elementos constitutivos de las proteínas. (Cap. 2, Sec. 1, pág. 37)

amniotic sac / saco amniótico: membrana protectora, delgada y llena de líquido, que se forma alrededor del embrión. (Cap. 6, Sec. 3, pág. 159)

antibody / anticuerpo: proteína que se produce como respuesta a un antígeno específico; se puede adherir al antígeno y lo puede anular. (Cap. 7, Sec. 1, pág. 178)

antigen / antígeno: molécula compleja extraña al cuerpo. (Cap. 7, Sec. 1, pág. 178)

artery / arteria: vaso sanguíneo que saca sangre del corazón y que tiene gruesas paredes elásticas hechas de tejido conectivo y tejido de músculo liso. (Cap. 3, Sec. 1, pág. 68)

asthma / asma: trastorno pulmonar en el cual los tubos bronquiales se contraen rápidamente y dificultan la respiración y causan estornudo o tos; puede ocurrir como una reacción alérgica. (Cap. 4, Sec. 1, pág. 100)

atriums / atrios: las dos cavidades superiores del corazón que se contraen al mismo tiempo durante un latido del corazón. (Cap. 3, Sec. 1, pág. 65)

axon / axón: estructura de la neurona que transmite mensajes de las células corporales. (Cap. 5, Sec. 1, pág. 119)

B

biological vector / vector biológico: organismo portador de enfermedad, como por ejemplo, una rata, un mosquito o una mosca, que propaga enfermedades contagiosas. (Cap. 7, Sec. 2, pág. 185)

bladder / vejiga: órgano elástico y muscular que retiene la orina hasta que ésta sale del cuerpo por la uretra. (Cap. 4, Sec. 2, pág. 104)

brain stem / bulbo raquídeo: conecta el encéfalo a la médula espinal y está compuesto del encéfalo medio, el puente de Varolio y la médula. (Cap. 5, Sec. 1, pág. 122)

bronchi / bronquios: dos conductos cortos que se bifurcan del extremo inferior de la tráquea y por los cuales se introduce el aire en los pulmones. (Cap. 4, Sec. 1, pág. 95)

C

capillary / capilar: vaso sanguíneo microscópico que conecta las arterias y las venas; tiene paredes de una célula de grosor, a través de las cuales se difunden los nutrientes y el oxígeno por todas las células corporales y se extraen los materiales de desecho y el dióxido de carbono. (Cap. 3, Sec. 1, pág. 69)

carbohydrates / carbohidratos: la fuente principal de energía del cuerpo, azúcares, almidones y fibra, formada de átomos de carbono, hidrógeno y oxígeno; se encuentra en las frutas, en las pastas y en los cereales y panes integrales. (Cap. 2, Sec. 1, pág. 38)

cardiac muscle / músculo cardíaco: músculo involuntario y estriado que sólo se encuentra en el corazón. (Cap. 1, Sec. 2, pág. 17)

cartilage / cartílago: resbaladiza capa de tejido flexible, gruesa y lisa que cubre los extremos de los huesos; facilita el movimiento al reducir la fricción y absorbe los choques. (Cap. 1, Sec. 1, pág. 10)

central nervous system / sistema nervioso central: división del sistema nervioso central, compuesta del encéfalo y la médula espinal. (Cap. 5, Sec. 1, pág. 121)

cerebellum / cerebelo: parte del encéfalo que controla los movimientos de los músculos voluntarios, mantienen el tono muscular y ayuda a mantener el equilibrio. (Cap. 5, Sec. 1, pág. 122)

cerebrum / cerebro: la parte más grande del encéfalo, donde se almacena la memoria, se controlan los movimientos y se interpretan los impulsos de los sentidos. (Cap. 5, Sec. 1, pág. 122)

chemical digestion / digestión química: proceso del sistema digestivo que se lleva a cabo cuando las sustancias químicas descomponen las moléculas grandes de alimentos en moléculas más pequeñas que pueden ser absorbidas por el cuerpo. (Cap. 2, Sec. 2, pág. 47)

chemotherapy / quimioterapia: uso de sustancias químicas para destruir células cancerosas. (Cap. 7, Sec. 3, pág. 194)

chyme / quimo: producto acuoso de la digestión que se mueve lentamente desde el estómago y entra en el intestino delgado. (Cap. 2, Sec. 2, pág. 51)

cochlea / cóclea: estructura llena de fluido en el oído interno en donde las vibraciones sonoras se convierten en impulsos nerviosos que son enviados al encéfalo. (Cap. 5, Sec. 2, pág. 132)

coronary circulation / circulación coronaria: flujo sanguíneo hacia los tejidos del corazón y fuera de éstos. (Cap. 3, Sec. 1, pág. 65)

D

dendrite / dendrita: estructura de la neurona que recibe mensajes y los envía a las células corporales. (Cap. 5, Sec. 1, pág. 119)

dermis / dermis: capa de la piel debajo de la epidermis que contiene vasos sanguíneos, nervios, glándulas sudoríparas y sebáceas y otras estructuras. (Cap. 1, Sec. 3, pág. 21)

diaphragm / diafragma: músculo situado debajo de los pulmones que se contrae y se relaja permitiendo así la entrada y salida de gases del cuerpo. (Cap. 4, Sec. 1, pág. 96)

digestion / digestión: proceso químico y mecánico que rompe los alimentos en partículas pequeñas de modo que el cuerpo las pueda absorber. (Cap. 2, Sec. 2, pág. 47)

E

embryo / embrión: óvulo fecundado adherido a la pared uterina. (Cap. 6, Sec. 3, pág. 159)

emphysema / enfisema: enfermedad pulmonar en la cual se produce una dilatación de los alvéolos. (Cap. 4, Sec. 1, pág. 99)

enzyme / enzima: tipo de proteína que acelera la velocidad de una reacción química, pero que no se agota o se cambia de ninguna forma durante la reacción. (Cap. 2, Sec. 2, pág. 48)

epidermis / epidermis: capa exterior de la piel y la más delgada, la cual produce constantemente nuevas células para reemplazar las células viejas que se desprenden de su superficie. (Cap. 1, Sec. 3, pág. 20)

F

fat / grasa: tipo de nutriente que provee energía y que ayuda al cuerpo a absorber vitaminas; las grasas pueden ser saturadas (carnes y quesos) o insaturadas (el aceite líquido y las grasas de carnes de aves de corral, pescado y nueces). (Cap. 2, Sec. 1, pág. 39)

fetal stress / estrés fetal: puede presentarse durante el proceso de alumbramiento o después del nacimiento conforme el lactante se ajusta de un entorno acuoso, oscuro y de temperatura constante a su nuevo entorno. (Cap. 6, Sec. 3, pág. 162)

fetus / feto: bebé en desarrollo después de los primeros dos meses de embarazo hasta su nacimiento. (Cap. 6, Sec. 3, pág. 160)

food group / grupo alimenticio: uno de los cinco grupos de alimentos que contiene los mismos nutrientes: panes y cereales, vegetales, frutas, leche y carnes. (Cap. 2, Sec. 1, pág. 44)

H

hemoglobin / hemoglobina: sustancia química en los glóbulos rojos que transporta oxígeno desde los pulmones hacia las células corporales y lleva parte del dióxido de carbono desde las células corporales, de regreso a los pulmones,. (Cap. 3, Sec. 2, pág. 75)

homeostasis / homeostasis: regulación de las condiciones internas de un organismo, las cuales lo mantienen vivo a pesar de los cambios en su ambiente. (Cap. 5, Sec. 1, pág. 119)

hormone / hormona: sustancia química que produce el sistema endocrino, se libera directamente en el torrente sanguíneo a través de glándulas sin conductos, actúa en tejidos asignados y puede acelerar o aminorar las actividades celulares. (Cap. 6, Sec. 1, pág. 146)

I

immune system / sistema inmunológico: grupo complejo de defensas que protege al cuerpo contra patógenos; este sistema incluye la piel y los sistemas respiratorio digestivo y circulatorio. (Cap. 7, Sec. 1, pág. 176)

infectious disease / enfermedad contagiosa: enfermedad causada por un virus, una bacteria, un hongo o un protista y que se propaga de un organismo infectado o del medio ambiente a otro organismo. (Cap. 7, Sec. 2, pág. 185)

involuntary muscle / músculo involuntario: músculo que no se puede controlar conscientemente, como el del corazón. (Cap. 1, Sec. 2, pág. 15)

J

joint / articulación: cualquier lugar en donde se unen dos o más huesos; puede ser móvil o fija. (Cap. 1, Sec. 1, pág. 11)

K

kidney / riñón: órgano del sistema urinario en forma de frijol y que está formado por cerca de 1 millón de nefrones; filtra la sangre produciendo la orina. (Cap. 4, Sec. 2, pág. 102)

L

larynx / laringe: vía respiratoria a la cual se encuentran adheridas las cuerdas vocales. (Cap. 4, Sec. 1, pág. 95)

ligament / ligamento: banda resistente de tejido que mantiene unidos los huesos en las articulaciones. (Cap. 1, Sec. 1, pág. 11)

lymph / linfa: fluido tisular que se difunde en los capilares. (Cap. 3, Sec. 3, pág. 80)

lymphatic system / sistema linfático: transporta la linfa a través de una red de capilares y vasos linfáticos y la vacía en venas grandes cerca del corazón; ayuda a combatir infecciones y enfermedades. (Cap. 3, Sec. 3, pág. 80)

lymph node / ganglio linfático: órgano en forma de frijol ubicado por todo el cuerpo y el cual filtra los microorganismos y materias foráneas que han encontrado los linfocitos. (Cap. 3, Sec. 3, pág. 81)

lymphocyte / linfocito: tipo de glóbulo blanco que combate las infecciones. (Cap. 3, Sec. 3, pág. 80)

M

mechanical digestion / digestión mecánica: proceso del sistema digestivo que rompe los alimentos al masticarlos, mezclarlos y revolverlos. (Cap. 2, Sec. 2, pág. 47)

melanin / melanina: pigmento producido por la epidermis que protege la piel del daño causado por el Sol y que le provee a la piel y a los ojos su color. (Cap. 1, Sec. 3, pág. 21)

menstrual cycle / ciclo menstrual: ciclo de cambios mensual controlado por hormonas del sistema reproductor femenino. Incluye la maduración de un óvulo y la preparación del útero para un posible embarazo. (Cap. 6, Sec. 2, pág. 154)

menstruation / menstruación: descarga mensual de sangre y células tisulares que ocurre cuando el revestimiento uterino se desintegra y se desprende. (Cap. 6, Sec. 2, pág. 154)

mineral / mineral: nutriente inorgánico que regula gran parte de las reacciones químicas del cuerpo; genera células, envía impulsos nerviosos y transporta oxígeno. (Cap. 2, Sec. 1, pág. 42)

muscle / músculo: órgano que se puede relajar, contraer y que provee la fuerza para mover los huesos y las partes del cuerpo. (Cap. 1, Sec. 2, pág. 14)

N

nephron / nefrón: diminuta unidad filtradora del riñón. (Cap. 4, Sec. 2, pág. 103)

neuron / neurona: unidad básica funcional del sistema nervioso; compuesta del cuerpo celular, dendritas y axones. (Cap. 5 Sec. 1, pág. 119)

noninfectious disease / enfermedad no contagiosa: enfermedad como, por ejemplo, el cáncer, la diabetes o el asma que no se propaga de una persona a otra. (Cap. 7, Sec. 3, pág. 190)

nutrients / nutrientes: sustancias en los alimentos (proteínas, carbohidratos, grasas, vitaminas, minerales y agua) que el cuerpo usa para el desarrollo, reparación y crecimiento celulares y que proveen energía. (Cap. 2, Sec. 1, pág. 36)

===== O =====

olfactory cell / célula olfativa: célula nerviosa nasal estimulada por moléculas en el aire, la cual envía impulsos al encéfalo, el cual interpreta los olores. (Cap. 5, Sec. 2, pág. 133)

ovary / ovario: órgano reproductor femenino que produce óvulo; se encuentra ubicado en la parte inferior del cuerpo. (Cap. 6, Sec. 2, pág. 153)

ovulation / ovulación: proceso mensual en que un ovario libera un óvulo que entra en el oviducto donde un espermatozoide puede fecundarlo. (Cap. 6, Sec. 2, pág. 153)

===== P =====

passive immunity / inmunidad pasiva: inmunidad que resulta cuando el cuerpo recibe anticuerpos producidos en otro animal; no dura tanto tiempo como la inmunidad activa. (Cap. 7, Sec. 1, pág. 179)

pasteurization / pasteurización: proceso que consiste en calentar un líquido a una temperatura que destruye la mayoría de las bacterias. (Cap. 7, Sec. 2, pág. 181)

periosteum / periósteo: membrana resistente y apretada que cubre la superficie de los huesos y que contiene vasos sanguíneos que transportan nutrientes al hueso. (Cap. 1, Sec. 1, pág. 9)

peripheral nervous system / sistema nervioso periférico: división del sistema nervioso; comprende todos los nervios fuera del sistema nervioso central, conecta el encéfalo y la médula espinal con otras partes del cuerpo. (Cap. 5, Sec. 1, pág. 121)

peristalsis / peristalsis: contracciones musculares que mueven los alimentos a través del sistema digestivo. (Cap. 2, Sec. 2, pág. 50)

pharynx / faringe: región en forma de conducto por donde pasan los alimentos, los líquidos y el aire. (Cap. 4, Sec. 1, pág. 94)

plasma / plasma: parte líquida de la sangre, cuyo constituyente principal es el agua, en el cual se disuelven el oxígeno, los nutrientes y los minerales. (Cap. 3, Sec. 2, pág. 74)

platelet / plaqueta: fragmento celular de forma irregular que facilita la coagulación de la sangre y libera sustancias químicas que ayudan a formar la fibrina. (Cap. 3, Sec. 2, pág. 75)

pregnancy / embarzo: período de desarrollo, generalmente unas 38 ó 39 semanas en los humanos, desde la fecundación del óvulo hasta el nacimiento. (Cap. 6, Sec. 3, pág. 158)

protein / proteína: tipo de nutriente compuesto de aminoácidos y el cual se usa para el reemplazo y la reparación de las células corporales, tiene moléculas grandes que contienen carbono, hidrógeno, oxígeno y nitrógeno; se encuentra en las carnes, pescado y frijoles. (Cap. 2, Sec. 1, pág. 37)

pulmonary circulation / circulación pulmonar: flujo de sangre que va del corazón a los pulmones y regresa al corazón. (Cap. 3, Sec. 1, pág. 66)

R

reflex / reflejo: respuesta involuntaria automática a un estímulo, controlada por la médula espinal. (Cap. 5, Sec. 1, pág. 125)

retina / retina: tejido sensible a la luz en la parte posterior del ojo; contiene bastones y conos. (Cap. 5, Sec. 2, pág. 129)

S

semen / semen: mezcla de espermatozoides y un líquido que ayuda a los espermatozoides a moverse y que les sirve como una fuente de energía. (Cap. 6, Sec. 2, pág. 152)

sexually transmitted disease (STD) / enfermedad transmitida sexualmente (ETS): enfermedad contagiosa, como la clamidia, el SIDA o el herpes genital, que se transmite de una persona a otra durante el contacto sexual. (Cap. 7, Sec. 2, pág. 186)

skeletal muscle / músculo óseo: músculo estriado voluntario que mueve los huesos; funciona en pares y se adhiere a los huesos mediante tendones. (Cap. 1, Sec. 2, pág. 17)

skeletal system / sistema óseo: todos los huesos del cuerpo; forma un marco vivo interno que provee forma y apoyo, protege los órganos internos, mueve los huesos, forma las células sanguíneas y almacena compuestos de calcio y fósforo para uso posterior. (Cap. 1, Sec. 1, pág. 8)

smooth muscle / músculo liso: músculo involuntario no estriado que controla el movimiento de los órganos internos. (Cap. 1, Sec. 2, pág. 17)

sperm / espermatozoides: células reproductoras masculinas que producen los testículos. (Cap. 6, Sec. 2, pág. 152)

synapse / sinapsis: pequeño espacio a través del cual se mueve un impulso desde un axón a las dendritas o al cuerpo celular de otra neurona. (Cap. 5, Sec. 1, pág. 121)

systemic circulation / circulación sistémica: la parte más grande del sistema circulatorio en que la sangre rica en oxígeno fluye a todos los órganos y tejidos corporales, excepto el corazón y los pulmones, y la sangre carente de oxígeno es devuelta al corazón. (Cap. 3, Sec. 1, pág. 67)

T

taste bud / papilas gustativas: principal receptor sensorial de la lengua; contiene vellos gustativos que envían impulsos al encéfalo, el cual interpreta los sabores. (Cap. 5, Sec. 2, pág. 134)

tendon / tendón: banda gruesa de tejido que adhiere los huesos a los músculos. (Cap. 1, Sec. 2, pág. 17)

testis / testículos: órgano masculino productor de espermatozoides y testosterona. (Cap. 6, Sec. 2, pág. 152)

trachea / tráquea: conducto transportador de aire que une la laringe con los bronquios, está forrado de membranas mucosas y cilios y contiene anillos cartilaginosos resistentes. (Cap. 4, Sec. 1, pág. 95)

U

ureter / uréter: conducto que transporta la orina desde los riñones hasta la vejiga. (Cap. 4, Sec. 2, pág. 104)

urethra / uretra: conducto que transporta la orina desde la vejiga y la expulsa del cuerpo. (Cap. 4, Sec. 2, pág. 104)

urinary system / sistema urinario: sistema de órganos excretores que elimina los residuos de la sangre, controla el volumen sanguíneo al eliminar el exceso de agua y mantiene el equilibrio en las concentraciones de sal y agua. (Cap. 4, Sec. 2, pág. 101)

urine / orina: líquido residual que contiene el exceso de agua, sales y otros residuos que el cuerpo no reabsorbe. (Cap. 4, Sec. 2, pág. 102)

uterus / útero: órgano hueco, muscular y con forma de pera donde un óvulo fecundado se desarrolla hasta convertirse en un bebé. (Cap. 6, Sec. 2, pág. 153)

V

vaccination / vacunación: acción y efecto de dar una vacuna a una persona o un animal, ya sea por vía oral o por inyección, con el fin de inmunizarlo contra una enfermedad. (Cap. 7, Sec. 1, pág. 179)

vagina / vagina: conducto muscular que conecta el extremo inferior del útero con la parte externa del cuerpo; el canal del nacimiento por el cual pasa el bebé cuando está naciendo. (Cap. 6, Sec. 2, pág. 153)

vein / vena: vaso sanguíneo que regresa la sangre al corazón y que posee válvulas de una sola vía que impiden que la sangre sea devuelta al corazón. (Cap. 3, Sec. 1, pág. 68)

ventricles / ventrículos: las dos cavidades inferiores del corazón que se contraen al mismo tiempo durante un latido del corazón. (Cap. 3, Sec. 1, pág. 65)

villi / vellosidades intestinales: prolongaciones que parecen dedos en las paredes del intestino delgado y que aumentan el área de superficie para la absorción de las moléculas de nutrientes. (Cap. 2, Sec. 2, pág. 52)

virus / virus: pequeñísimo trozo de material genético que infecta y se multiplica en células huéspedes. (Cap. 7, Sec. 2, pág. 182)

vitamin / vitamina: nutriente orgánico hidrosoluble y liposoluble necesario en pequeñas cantidades para el crecimiento, la regulación de las funciones corporales y para la prevención de enfermedades. (Cap. 2, Sec. 1, pág. 40)

voluntary muscle / músculo voluntario: músculo que puede controlarse conscientemente, como por ejemplo, un músculo de una pierna o un brazo. (Cap. 1, Sec. 2, pág. 15)

The index for *Human Body Systems* will help you locate major topics in the book quickly and easily. Each entry in the index is followed by the number of the pages on which the entry is discussed. A page number given in boldfaced type indicates the page on which that entry is defined. A page number given in italic type indicates a page on which the entry is used in an illustration or photograph. The abbreviation *act.* indicates a page on which the entry is used in an activity.

A

Abdominal thrusts, 96, *97,* *act.* 108–109
Active immunity, 179
Active transport, 64, *64*
Activities, 25, 26–27, 46, 54–55, 73, 82–83, 107, 108–109, 127, 136–137, 156, 166–167, 189, 196–197
Adolescence, 162, 164
Adulthood, 162, 164–165, *164, 165*
AIDS, 187
Air: oxygen in, 92, *92*
Air pollution: and cancer, 195
Albumin, 105
Alcohol, 126
Allergens, 191, *191*
Allergies, 190–191, *190, 191*
Alveoli, *94,* **95,** *95*
Amino acids, 37
Amniotic sac, 159
Anemia, 79
Antibiotics, 186
Antibodies, 178, *178,* 179; in blood, 77, 78
Antigens, 77, **178**
Antihistamines, 191
Antiseptics, 184, *184*
Anus, 53
Anvil, 131, *131*
Aorta, 66
Appendices. *see* Reference Handbooks
Arteries, 68, *68*
Asbestos, 192, *192*
Asthma, 100, 191
Astronomy Integration, 130
Atherosclerosis, *70,* 71

Atmosphere: oxygen in, 92, *92*
Atrium, 65, *66*
Autonomic nervous system, 123
Axon, 119, *119*

B

Bacteria: in digestion, 53; immune system and, *176, 177;* infectious diseases and, 182, 186, *186;* reproduction rates of, 179; sexually transmitted diseases caused by, 186, *186;* tetanus and, 179, *180*
Balance, 132, *132*
Ball-and-socket joint, 12, *12*
Before You Read, 7, 35, 63, 91, 117, 145, 175
Bicarbonate, 52
Bile, 52, 105
Biological vectors, 185, *185*
Birth(s): development before, 158–160, *159, 160;* multiple, 158, *158;* process of, 160–161, *161;* stages after, 162–166, *162, 163, 164, 165*
Birth canal (vagina), 153, *153, 161*
Bladder, *103,* **104,** 105
Blood, 74–79; clotting of, 76, *76;* diseases of, 79, *79;* functions of, 74; parts of, 74–75, *74, 75;* transfusion of, 77, 78, *act.* 82–83
Blood cells: red, 75, *75, 76,* 79, *79;* white, 10, 75, *75, 76,* 79, 177, *177,* 193
Blood pressure, 69, 71, *71*
Blood types, 77–78, *act.* 82–83
Blood vessels, 64, *64,* 68–69; aorta, 66; arteries, 68, *68;*

bruises and, 23; capillaries, *68,* 69, 177, *177;* in regulation of body temperature, 22; veins, 68, *68*
Body: proportions of, 164, *164, act.* 166–167
Body temperature, 22, *22,* 177
Bone(s), 8–13, *9, 10;* compact, 9, *9;* formation of, 10, *10;* spongy, 10; structure of, 9–10, *9, act.* 26–27
Bone marrow, 10, 79
Brain, 122, *122*
Brain stem, 122, *122*
Breast cancer, 193
Breathing, 96, *96;* rate of, *act.* 91; respiration and, 92–93, *92*
Bronchi, 95
Bronchitis, 98, 99
Bruises, 23, *23*
Burns, 24
Caffeine, 126, *126*

C

Calcium, 42; in bones, 10
Calcium phosphate, 9
Cancer, 98, 100, *100,* 185, 193–195; causes of, 192, *192,* 194, *194;* early warning signs of, 195; prevention of, 195; treatment of, 194
Capillaries, 68, **69,** *94,* 95, *95,* 177, *177*
Carbohydrates, 38, *38*
Carbon dioxide: as waste product, 93, 95, 96
Carcinogen(s), 100, 194, *194*
Cardiac muscles, 17, *17*
Cardiovascular disease, *70,* 71–72

Cardiovascular system, 64–73; blood pressure and, 69, 71, *71;* blood vessels in, 64, *64,* 66, 68–69, *68;* diffusion in, 64, *64;* heart in, 65–67, *65,* act. 73

Career, 2–3, 28, 56–57, 85, 110–111, 139

Cartilage, 10, *10,* 13, *13*

Cavities: in bone, 10

Cell(s): cancer, 193; division, 157; T-cells, 178, *178,* 194

Cellular respiration, 93

Centers for Disease Control and Prevention (CDC), 185

Central nervous system, 121–123, *121*

Cerebellum, 122, *122*

Cerebral cortex, *119,* 122

Cerebrum, 122, *122*

Cervix, *153*

Cesarean section, 161

Chemical digestion, 47

Chemical messages: hormones and, 146; modeling, *act.* 145

Chemicals: and disease, 192, *192*

Chemistry Integration, 21, 78, 122

Chemotherapy, 194

Chicken pox, 179

Childbirth. *see* Birth(s)

Childhood, 162, 163, *163, 164*

Chlamydia, 186

Choking: abdominal thrusts for, 96, *97, act.* 108–109

Cholesterol, 39, 71

Chronic bronchitis, 98, 99

Chronic diseases, 190–191, *190, 191*

Chyme, 51

Cilia, *90,* 94, *94,* 98, 153

Circulation: coronary, **65,** *65;* pulmonary, **66,** *66;* systemic, **67,** *67*

Circulatory system, 62, *62, act.* 63, 64–73; blood pressure and, 69, 71, *71;* blood vessels in, 64, *64,* 66, 68–69, *68;* heart in, 65–67, *65, act.* 73; pathogens and, 177, *177*

Classification: of joints, 12; of muscle tissue, 18, *18*

Clean Air Act, 195

Cleanliness: and infectious diseases, 184, *184,* 188, *188*

Clotting, 76, *76*

Cochlea, *131,* **132**

Cold virus, 98

Colorectal cancer, 193

Compact bone, 9, *9*

Complex carbohydrates, 38

Concave lens, 129, *129,* 130, *130*

Cones, 129

Control, 116, *act.* 117

Convex lens, 129, *129,* 130, *130*

Coordination, 116

Coronary circulation, 65, *65*

Cortex, *119,* 122

Cristae ampullaris, 132, *132*

Cuts, 23

Cycles: menstrual, 154–155, *154*

D

Dendrite, 119, *119*

Depressant, 126

Dermis, *20,* **21,** 23

Design Your Own Experiment, 82–83, 136–137, 196–197

Developmental stages, 162–165, *162, 163, 164, 165*

Diabetes, 105, 185, 191, *191*

Diagrams: interpreting, *act.* 156

Dialysis, 106, *106*

Diaphragm, 96

Diastolic pressure, 69

Diffusion: in cardiovascular system, 64, *64*

Digestion, 47; bacteria in, 53; chemical, **47;** enzymes in, 48, 50, 51, 52; food particle size and, *act.* 54–55; mechanical, **47**

Digestive system: functions of, 47, *47;* human, 34, *34, act.* 35, 47–55, *49;* immune defenses of, 177; organs of, 49–53, *49*

Dinosaurs, 6, *6*

Dioxin, 193

Diphtheria, 179, *180*

Diseases: chemicals and, 192, *192;* chronic, 190–191, *190, 191;* cleanliness and, 184, *184,* 188, *188;* fighting, 188, *188;* infectious, 181–188, **185;** noninfectious, **190–195;** percentage of deaths due to, 185; of respiratory system, 98–100, *99, 100;* sexually transmitted, **186–187,** *186, 187;* spread of, *act.* 175, 185, *185;* of urinary system, 105–106, *106*

Disks, 13, *13*

Drugs: and nervous system, 126, *126*

Ducts, 147

Duodenum, *51,* 52

E

Ear, 131–132, *131, 132*

Eardrum, 131, *131*

Earth Science Integration, 23, 43, 93, 105, 147, 182

Eggs, 153, 155, *156,* 157, *157*

Embryo, 159, *159*

Emphysema, 99, *99*

Endocrine glands, 147, *148–149*

Endocrine system, 146–150; functions of, 146, *146;* menstrual cycle and, 154; reproductive system and, 151, *151*

Energy: muscle activity and, 19, *19;* nutrition and, 36

Environmental Science Integration, 53, 193

Enzymes, 48, *48;* chemical reactions and, 49; in digestion, 48, 50, 51, 52; pathogens and, 177

Epidermis, 20, *20,* 22

Epiglottis, 50, 94

Esophagus, 49

Estrogen, 154

Ethyl alcohol, 192

Excretory system, 101–107; diseases and disorders of, 105–106, *106;* functions of, 101; urinary system, 101–104, *101, 102, 103*

Exhaling, 96, *96*
Explore Activity, 7, 35, 63, 91, 117, 145, 175
Extensor muscles, *18*
Eye, 128–130, *128, 129, 130*

F

Farsightedness, 130, *130*
Fat(s), **39,** *39;* body, 72; dietary, 71
Feces, 53
Female reproductive system, 153–155, *153, act.* 156
Fertilization, 153, 155, 157, *157, 159*
Fetal stress, **162**
Fetus, 160, *160*
Fiber, 38
Fibrin, 76, *76*
Field Guides: Emergencies Field Guide, 208–211
Filtration: in kidneys, 103, *103, act.* 107
Flexor muscles, *18*
Flu, 181, 185
Fluid levels: regulation of, 102, *102,* 104, *105*
Focal point, 129, *129*
Foldables, 7, 35, 63, 91, 117, 145, 175
Food groups, **44**–45, *44*
Food labels, 45, *45*
Fraternal twins, 158, *158*
Fulcrum, 16
Fungi: and infectious diseases, 182

G

Gallbladder, 49, *49*
Glenn, John, *165*
Gliding joint, 12, *12*
Glucose: 52; calculating percentage in blood, 147; diabetes and, 191, *191*
Gonorrhea, 186, *186*
Growth spurt, 164

H

Hair follicle, *20*
Hammer, 131, *131*
Hamstring muscles, *18*
Handbooks. *see* Math Skill Handbook, Reference Handbook, Science Skill Handbook, and Technology Skill Handbook
Haversian systems, *9*
Hearing, 131–132, *131, 132*
Heart, 65–67, *65, act.* 73
Heart attack, 71
Heart disease, *70,* 71–72, 98, 185
Heart failure, 71
Heat transfer: in body, 22
Helper T-cells, 81
Hemoglobin, 23, **75**
Hemophilia, 76
Herpes, 186
High blood pressure, 71
Hinge joint, 12, *12*
Histamines, 191
HIV, 81, 187, *187*
Hives, *190*
Homeostasis, 53, **119,** 125, 126, 135
Hormones, **146;** menstrual cycle and, 154, 155; regulation of, 146–150, *150;* reproductive system and, 151, *151*
Human immunodeficiency virus. *see* HIV
Hydrochloric acid, 51
Hypertension, 71
Hypothalamus, 102, 154

I

Identical twins, 158, *158*
Immune system, **176**–180; antibodies and, 178, *178,* 179; antigens and, 178; first-line defenses in, 176–177, *176, 177, act.* 196–197; human immunodeficiency virus (HIV) and, 187, *187;* inflammation and, 177; specific immunity and, 178, *178*

Immunity, active, **179;** passive, **179,** 180, *180;* specific, 178, *178*
Immunology, 194
Infancy, 162–163, *162, 163*
Infections: and lymphatic system, 81
Infectious diseases, 181–188, **185;** cleanliness and, 184, *184,* 188, *188;* fighting, 188, *188;* in history, 181–184; Koch's rules and, 182, *183;* microorganisms and, 182, 186–187, *186, 187, act.* 189; sexually transmitted, **186**–187, *186, 187;* spread of, *act.* 175, 185, *185*
Inferior vena cava, 68
Inflammation, 177
Influenza, 181, 185
Inhaling, 96, *96*
Inner ear, *131,* 132, *132*
Insulin, 191
Interneuron, 119, *120*
Intestines: large, 53; small, 52, *52*
Involuntary muscles, 15, *15*
Iodine, 42, 147
Iron, 42

J

Joints, 11–12, *12*

K

Kidney(s), **102;** dialysis and, 106, *106;* diseases affecting, 105–106, *106;* filtration in, 103, *103, act.* 107; in regulation of fluid levels, 102, *102;* structure of, *act.* 107; transplantation of, 110–111
Koch, Robert, 182
Koch's rules, 182, *183*

L

Lab(s). *see* Activities, MiniLABs, and Try at Home MiniLABs
Labeling: of foods, 45, *45*

Index

Index

Large intestine, 53
Larynx, 95
Lenses, 129–130, *129, 130*
Leukemia, 79, 193
Levers, 16
Life span: human, 165
Ligament, 11
Lipids, 39, *39*
Lister, Joseph, 184
Liver, 49, *49,* 52
Lung(s), *94,* 95, 96, *96;* diseases of, 98–100, *99, 100*
Lung cancer, 98, 100, *100,* 192, 193, 194
Lymph, 80
Lymphatic system, 80–81, *80*
Lymph nodes, 81
Lymphocytes, 80, 81, 178, *178*

M

Maculae, 132, *132*
Malaria, 182
Male reproductive system, 150, *150*
Mammals: skeletal systems of, *act.* 26–27
Math Skill Handbook, 230–236
Math Skills Activities, 11, 133, 147
Measles, 179
Mechanical digestion, 47
Medulla, 122
Melanin, 21, *21*
Menopause, 155, *155*
Menstrual cycle, 154–155, *154*
Menstruation, 154–155, *154*
Microorganisms: and infectious diseases, 182, *182,* 186–187, *186, 187, act.* 189
Midbrain, 122
Middle ear, 131, *131*
Minerals, 42, *42*
MiniLABs: Recognizing Why You Sweat, 22; Comparing the Fat Content of Foods, 39; Modeling Scab Formation, 76; Modeling Kidney Function, 103; Comparing Sense of Smell, 134; Graphing Hormone Levels, 154; Observing Antiseptic Action, 184

Morula stage of development, *144*
Model and Invent, 108–109
Motor neuron, 119, *120*
Mouth: digestion in, 50, *50*
Movable joints, 12, *12*
Movement: cartilage and, 13, *13;* joints and, 11–12, *12;* levers and, 15, *16;* muscles and, *act.* 7, *15,* 18–19, *18, 19*
Mucus, 177
Multiple births, 158, *158*
Mumps, 179
Muscles, 14–19, *14;* cardiac, **17,** *17;* changes in, 18; classification of, 17, *17;* control of, 15, *15;* energy and,19, *19;* involuntary, **15,** *15;* movement and, *act.* 7, *14,* 18–19, *18, 19;* skeletal, **17,** *17;* smooth, **17,** *17;* voluntary, **15,** *15*

N

National Geographic Visualizing: Human Body Levers, 16; Vitamins, 41; Atherosclerosis, 70; Abdominal Thrusts, 97; Nerve Impulse Pathways, 120; The Endocrine System, 148; Koch's Rules, 183
Navel, 161
Nearsightedness, 130, *130*
Negative-feedback system, 150, *150*
Neonatal period, 162
Nephron, 103, *103*
Nerve cells, 119, *119*
Nervous system, 118–127; autonomic, 123; brain in, 122, *122;* central, **121**–123, *121;* drugs and, 126, *126;* injury to, 124, *124;* neurons in, 119, *119, 120,* 121, *121;* peripheral, **121,** 123; reaction time and, *act.* 127; reflexes and, 125, *125;* responses of, *act.* 117, 118–119, 125, *125;* safety and, 124–125; somatic,

123; spinal cord in, 123, *123,* 124, *124;* synapses in, 121, *121*
Neuron, 119, *119, 120,* 121, *121*
Niacin, 53
Noninfectious diseases, 190–195; cancer, 185, 193–195; chemicals and, 192, *192*
Nutrients, 36, 37–44; carbohydrates, 38, *38;* fats, 39, *39;* minerals, 42, *42;* proteins, 37, *37;* vitamins, 40, *41, act.* 46, 53; water, 43–44, *43*
Nutrition, 36–46, *36;* anemia and, 79; cancer and, 195; energy needs and, 36; food groups and, 44–45, *44;* heart disease and, 71, 72

O

Oil glands, *20,* 21
Older adulthood, 165, *165*
Olfactory cells, 133
Oops: Accidents in Science, 28–29
Osteoblasts, 10
Osteoclasts, 10
Outer ear, 131, *131*
Ovary, 153, *153,* 155, *156*
Oviduct, *153, 156,* 157
Ovulation, 153, 155
Oxygen: and respiration, 92, *92,* 95
Pain, 135

P

Pancreas, 49, *49,* 52
Paralysis, 124, *124*
Passive immunity, 179, 180, *180*
Pasteur, Louis, 181
Pasteurization, 181
Pathogens: and immune system, 176–177, *176, 177, 178,* 179
Periosteum, 9
Peripheral nervous system, 121, 123
Peristalsis, 50, 51, 52, 53
Pertussis, 179

Pharynx, 94
Phosphorus, 42; in bones, 10
Physics Integration, 69, 164
Pituitary gland, 151, *151*
Pivot joint, 12, *12*
Placenta, 159
Plasma, 74, 77
Platelets, 75, *75, 76, 76*
Pneumonia, 98, 185
Pollution: of air, 195
Pons, 122
Potassium, 42
Pregnancy, 158–160, *159*
Problem-Solving Activities, 40, 78, 104, 185
Progesterone, 154
Prostate cancer, 193
Proteins, 37, *37*
Protist(s), 182, *182*
Puberty, 164
Pulmonary circulation, 66, *66*

Quadriceps, *18*

Radiation: as cancer treatment, 194
Reaction time, improving, *act.* 127
Rectum, 53
Reference Handbook, 237–243
Reflex, 125, *125*
Regulation, 144, *144. see also* Endocrine system; chemical messages in, *act.* 145; of hormones, 146–150, *150*
Reproductive system, 151–156; endocrine system and, 151, *151;* female, 153–155, *153, act.* 156; function of, 157; hormones and, 151, *151;* male, 152, *152*
Respiration, 90*;* cellular, 93; oxygen and, 92, *92,* 95
Respiratory infections, 98
Respiratory system, 91–100; diseases and disorders of,

98–100, *99, 100;* functions of, 92–93, *93;* immune defenses of, 177; organs of, 94–95, *94*
Responses, *act.* 117, 118–119, 125, *125*
Retina, 129, 130
Rh factor, 78
Rods, 129
Rubella, 179

Safety: and nervous system, 124–125
Saliva, 50, *50,* 134, *act.* 196–197
Salivary glands, 49, *49,* 50, *50*
Saturated fats, 39
Scab, 23, 76
Science and Language Arts, 138–139
Science and History. *see* TIME
Science and Society. *see* TIME
Science Online: Data Update, 81, 187; Research, 10, 15, 38, 50, 71, 75, 81, 95, 98, 123, 125, 133, 153, 161, 178
Science Skill Handbook, 212–225
Science Stats, 168–169, 198–199
Scrotum, 152, *152*
Semen, 152
Semicircular canals, 132, *132*
Seminal vesicle, 152, *152*
Senses, 128–137; as alert system, 128; hearing, 131–132, *131, 132;* skin sensitivity, 135, *135, act.* 136–137; smell, 133–134; taste, 134, *134;* touch, 135, *135;* vision, 128–130, *128, 129, 130*
Sensory neuron, 119, *120*
Sensory receptors, 135, *135*
Sexually transmitted diseases (STDs), 186–187, *186, 187*
Sickle-cell anemia, 79, *79*
Skeletal muscles, 17, *17*
Skeletal system, 8–13, *9;* bones in, 8–10, *9, 10, act.* 26–27; cartilage in, 10, *10,* 13, *13;*

functions of, 8, *8;* joints in, 11–12, *12;* of mammals, *act.* 26–27
Skin, 20–24; functions of, 21–22, *22;* in immune system, 176, *177;* injuries to, 23, *23;* repairing, 23, 24, *24;* sensitivity of, 135, *135, act.* 136–137; structures of, 20–21, *20, 21*
Skin grafts, 23–24, *24*
Small intestine, 52, *52*
Smell, 133–134
Smoking, 194; cardiovascular disease and, 72, *72;* respiratory disease and, 98, 100, *100*
Smooth muscles, 17, *17*
Sodium, 42
Somatic nervous system, 123
Sound waves, 131, *131*
Specific immunity, 178, *178*
Sperm, 152, *152,* 155, 157, *157*
Spinal cord, 123, *123,* 124, *124*
Spinal nerves, 123, *123*
Spleen, 81
Spongy bone, 10
Standardized Test Practice, 33, 61, 89, 115, 143, 173, 203
Staphylococci bacteria, *177*
Starch, 38
Stimuli: responses to, *act.* 117, 118, 125
Stirrup, 131, *131,* 132
Stomach, 51, *51*
Striated muscles, 17, *17*
Stroke, 185
Structure. *see* Muscles; Skeletal system
Sugars, 38; in blood, 147. *see also* Glucose
Sunscreen lotion, 195
Superior vena cava, 68
Surface area, 96
Sweat glands, 20, 21, 22
Synapse, 121, *121*
Syphilis, 186, *186*

Index

Systemic circulation, 67, *67*
Systolic pressure, 69

T

Taste, 134, *134*
Taste buds, 134, *134*
T-cells, 81, 178, *178,* 194
Technology Skill Handbook, 226–229
Temperature: of human body, 22, *22;* pathogens and, 177
Tendons, 17
Test Practice. *see* Standardized Test Practice
Testes, 152, *152*
Testosterone, 152
Tetanus, 179, 180, *180*
The Princeton Review. *see* Standardized Test Practice
Thiamine, 53
Thyroid gland, 147
TIME: Science and History, 84–85, 110–111; Science and Society, 56–57
Tonsils, *176*
Touch, 135, *135*
Toxins, 192
Trachea, *94,* **95**
Traditional Activities, 54–55, 166–167

Transport: active, 64, *64*
Try at Home MiniLABs: Comparing Muscle Activity, 18; Modeling Absorption in the Small Intestine, 52; Inferring How Hard the Heart Works, 65; Comparing Surface Area, 96; Observing Balance Control, 132; Interpreting Fetal Development, 160; Determining Reproduction Rates, 179
Tuberculosis, 185
Tumor, 193
Twins, 158, *158*

U

Umbilical cord, 159, 161
Urea, 105
Ureter, 104
Urethra, *103,* **104,** 152, *152*
Urinary system, 101–104, *101;* diseases and disorders of, 105–106, *106;* organs of, 102–104, *103;* regulation of fluid levels by, 102, *102,* 104
Urine, 102, *102*
Use the Internet, 26–27
Uterus, 153, *153,* 154, *154, 156,* 158, 159, *159*

Vaccination, 179, 180
Vaccine, 179
Vagina (birth canal), 153, *153, 161*
Veins, 68, *68*
Ventricles, 65, *66,* 68
Vertebra, 13, *13,* 123, 124
Villi, 52, *52*
Virus(es), 182; sexually transmitted diseases caused by, 186–187, *187*
Vision, 128–130, *128, 129, 130*
Visualizing. *see* National Geographic
Vitamin(s), 40, *41;* **B,** 53; **C,** *act.* 46; **D,** 22; **K,** 53
Vocal cords, *94,* 95
Voluntary muscles, 15, *15*

W

Water: as nutrient, 43–44, *43*
Wave(s): sound, 131, *131*
White blood cells, 10, 177, *177,* 193
Whooping cough, 179
Withdrawal reflex, 125

Z

Zygote, 157, 158, 159, *159*

Index

Art Credits

Glencoe would like to acknowledge the artists and agencies who participated in illustrating this program: Absolute Science Illustration; Andrew Evansen; Argosy; Articulate Graphics; Craig Attebery represented by Frank & Jeff Lavaty; CHK America; Gagliano Graphics; Pedro Julio Gonzalez represented by Melissa Turk & The Artist Network; Robert Hynes represented by Mendola Ltd.; Morgan Cain & Associates; JTH Illustration; Laurie O'Keefe; Matthew Pippin represented by Beranbaum Artist's Representative; Precision Graphics; Publisher's Art; Rolin Graphics, Inc.; Wendy Smith represented by Melissa Turk & The Artist Network; Kevin Torline represented by Berendsen and Associates, Inc.; WILDlife ART; Phil Wilson represented by Cliff Knecht Artist Representative; Zoo Botanica.

Photo Credits

Abbreviation key: AA=Animals Animals; AH=Aaron Haupt; AMP=Amanita Pictures; BC=Bruce Coleman, Inc.; CB=CORBIS; DM=Doug Martin; DRK=DRK Photo; ES=Earth Scenes; FP=Fundamental Photographs; GH=Grant Heilman Photography; IC=Icon Images; KS=KS Studios; LA=Liaison Agency; MB=Mark Burnett; MM=Matt Meadows; PE=PhotoEdit; PD=PhotoDisc; PQ=PictureQuest; PR=Photo Researchers; SB=Stock Boston; TSA=Tom Stack & Associates; TSM=The Stock Market; VU=Visuals Unlimited.

Cover Dan McCoy/Rainbow; **v** AH; **vi** KS; **1**Oliver Meckes/PR; **2** Bettmann/CB; **3** (b)DOE Human Genome Program www.ornl.gov/hgmis, (t)AFP/CB; **4** (b)Raphael Gaillarde/LA, (t)Rosenfeld Images Ltd./Science Photo Library/PR; **6** PR; **6-7** Chris Butler/Science Photo Library/PR; **7** MM;

8 Michael St. Maur Sheil/CB; **9** John Serro/VU; **12** (t)PR, (b)Geoff Butler; **14** MB; **15** AH; **16** M. McCarron; **17** (l)Breck P. Kent, (c)Runk/Schoenberger from GH, (r)Carolina Biological Supply/PhotoTake NYC; **21** (t, l to r)Clyde H. Smith/Peter Arnold, Inc., Erik Sampers/PR, Michael A. Keller/TSM, Dean Conger/CB, (b, l to r)Ed Bock/TSM, Joe McDonald/VU, Art Stein/PR, Peter Turnley/CB; **23** Jim Grace/PR; **24** PR; **25** MB; **28 29** Sara Davis/The Herald-Sun; **30** (t)Ken Lucas/VU, (c)Steve Prezant/TSM, (b)Bob Krist/CB; **32** (tl)Breck P. Kent, (tr)Carolina Biological Supply/PhotoTake NYC, (b)Runk/Schoenberger from GH; **34** SIU/VU; **34-35** Meckes/Ottawa/PR; **36 37 38** KS; **39** (l)KS, (r)VU; **41** (nervous system)Sally J. Bensusen/Visual Science Studio, (cabbages, avocado)Artville, (skeleton, liver)DK Images, (doctor)Michael W. Thomas, (girls)Digital Vision/PQ, (cell membrane)Precision Graphics, (blood cells, blood clot)David M. Philips/VU, (others)Digital Stock; **42** Gary Kreyer from GH; **43** Larry Stepanowicz/VU; **44 45** KS; **47** (l)KS, (r)Tom McHugh/PR; **49** Geoff Butler; **51** (t)Dr. K.F.R. Schiller/PR, (c)Custom Medical Stock Photo, (b)Dr. K.F.R. Schiller/PR; **52** Biophoto Associates/PR; **54** (t)KS, (b)MM; **55** MM; **56** Eising/Stock Food; **57** Goldwater/Network/Saba; **58 59** KS; **62** Prof. S.H.E. Kaufmann & Dr. J.R. Golecki/Science Photo Library/PR; **62-63** Duncan Wherrett/Stone; **63** MB; **64 67** AH; **69** MM; **70** (tl cr br)Stephen R. Wagner, (others) Martin M. Rotker; **71** (t)StudiOhio, (b)MM; **73** First Image; **75** National Cancer Institute/Science Photo Library/PR; **79** Meckes/Ottawa/PR; **80** AH; **82** (t)MM/Peter Arnold, Inc., (b)MM; **85** (l)Science Photo Library/PR, (r)North Wind Picture Archive; **86** David Phillips/PR; **87** (tl) Manfred Kage/Peter Arnold, Inc., (tr)K.G. Murti/VU, (b)Don W. Fawcett/VU; **90** Professor P. Motta/Dept. of Anatomy/University "La Sapienza," Rome/Science Library/PR; **90-91** David Madison/Newsport Photography; **91** Geoff Butler; **92** Randy

Credits

Lincks/CB; **93** Dominic Oldershaw; **94** Bob Daemmrich; **97** Richard T. Nowitz; **99** (l c)SIU/PR, (r)Geoff Butler; **100** Renee Lynn/PR; **102** (l)Science Pictures Ltd./Science Photo Library/PR, (r)SIU/PR; **104** Paul Barton/TSM; **105** (l)Gunther/Explorer/PR, (c r)MB; **106** Richard Hutchings/PR; **107** Biophoto Associates/Science Source/PR; **108** (t)Larry Mulvehill/PR, (b)MM; **109** MM; **110** Lane Medical Library; **110-111** SPL/CB; **111** Custom Medical Stock Photo; **112** (t)Cabisco/VU, (b)Custom Medical Stock Photo; **113** (t)Gregg Ozzo/VU, (c)Ed Beck/TSM, (b)Tom & DeeAnn McCarthy; **116** Professor P. Motta/Dept. of Anatomy/University "La Sapienza," Rome/Science Photo Library/PR; **116-117** John Terrance Turner/FPG; **117** MM; **118** KS; **119** A. & F. Michler; **125** KS; **126** Michael Newman/PE; **127** KS; **131** AH; **135** MB; **136** (t)Jeff Greenberg/PE, (b)AMP; **137** AMP; **138** Toni Morrison; **139** Nic Paget-Clarke; **140** (tl)Bob Daemmrich/SB, (tr)MM, (b)Christel Rosenfeld/Stone; **141** (l)David R. Frazier/PR, (r)Michael Brennan/CB; **144** Profs. P.M. Motta & J. Van Blerkom/Science Photo Library/PR; **144-145** Brownie Harris/TSM; **145** John Evans; **146** David Young-Wolff/PE; **149** Stephen R. Wagner; **155** Ariel Skelley/TSM; **157** David M. Phillips/PR; **158** (l)Tim Davis/PR, (r)Chris Sorensen/TSM; **159** Science Pictures Ltd./Science Photo Library/PR; **160** Petit Format/Nestle/Science Source/PR; **162** (l)Jeffery W. Myers/SB, (r)Ruth Dixon; **163** (tl)MB, (tr)AH, (b)MB; **164** KS; **165** (l)NASA/Roger Ressmeyer/CB, (r)AFP/CB; **166** (t)Chris Carroll/CB, (b)Richard Hutchings; **167** MM; **168** (l)John Banagan/The Image Bank, (c)Ron Kimball Photography, (r)SuperStock; **169** (l)Joe McDonald/CB, (r)Martin B. Withers/Frank Lane Picture Agency/CB; **170** (t)DM, (c)David Woods/TSM, (b)David M. Phillips/PR; **171** (l)Bob Daemmrich, (r)Maria Taglienti/The Image Bank; **174** Oliver Meckes/PR; **174-175** S. Lowry/University Ulster/Stone; **175** KS; **176** Dr. P. Marazzi/Science Photo Library/PR; **177** (tl)Michael A. Keller/TSM, (tr)Runk/Schoenberger from GH, (b)NIBSC/Science Photo Library/PR; **180** CC Studio/Science Photo Library/PR; **183** (tl)VU, (tr)Jack Bostrack/VU, (cl)Cytographics Inc./VU, (cr)Cabisco/VU, (bl)Jack Bostrack/VU, (br)VU; **184** MM/Michelle Del Guercio/PR; **185** Nigel Cattlin, Holt Studios/PR; **186** (t)Oliver Meckes/Eye of Science/PR, (b)VU; **187** Oliver Meckes/Eye of Science/Gelderblom/PR; **188** MB; **190** (l)Caliendo/Custom Medical Stock Photo, (r)AMP; **191** (t)Andrew Syred/Science Photo Library/PR, (b)Custom Medical Stock Photo; **192** (t)Jan Stromme/BC, (c)J.Chiasson-Liats/LA, (b)Mug Shots/TSM; **196** (t)Tim Courlas, (b)MM; **197** MM; **198** Layne Kennedy/CB; **199** MB; **200** (t)Amethyst/Custom Medical Stock Photo, (c)Dr. P. Marazzi/Science Photo Library/PR, (b)USDA/Science Source/PR; **201** (tl)Gelderblom/Eye of Science/PR, (tr)Garry T. Cole/BPS/Stone, (bl)Oliver Meckes/Ottawa/PR, (br)David M. Philips/VU; **206-207** PD; **208** (t)Michael Newman/PE, (b)MAK 1 Photodesign; **209** (t)David Young-Wolff/PE, (b)W. Hill Jr./The Image Works; **210** (t)AH, (b)Marc Romanelli/The Image Bank; **211** (t)Rob Crandall/The Image Works, (c)AH, (b)Michael Newman/PE; **212** Timothy Fuller; **216** First Image; **219** Dominic Oldershaw; **220** StudiOhio; **221** First Image; **223** Richard Day/AA; **226** Paul Barton/TSM; **229** Charles Gupton/TSM; **237** MM; **240** (t)NIBSC/Science Photo Library/PR, (bl)Dr. Richard Kessel, (br)David John/VU; **241** (t)Runk/Schoenberger from GH, (bl)Andrew Syred/Science Photo Library/PR, (br)Rich Brommer; **242** (t)G.R. Roberts, (bl)Ralph Reinhold/ES, (br)Scott Johnson/AA; **243** Martin Harvey/DRK.

Acknowledgements

From *Sula*, by Toni Morrison. Copyright © 1973 by Toni Morrison. Reprinted by permission of International Creative Management, Inc.